Handbook of Model Rocketry

Fourth Edition

Completely Revised

Handbook of

MODEL

ROCKETRY

G. Harry Stine

Founder and past president,
National Association
of Rocketry

FOLLETT PUBLISHING COMPANY
Chicago

To: William S. Roe
Orville H. Carlisle
and a couple of thousand
young model rocketeers

Appreciation is expressed to the following
persons and organizations for permission to
reproduce photographs: L. Audin, page 204;
Al Kniele, page 125; Tom Pastrick, pages 12
and 203; O. Saffek, pages 275, 277, 278, and
327; Don Sahlin, pages 236 and 238; *Arizona
Journal,* page 173; Centuri Engineering
Company, pages 17, 98, 231, 260, 270, and
281; Estes Industries, Inc., pages 94, 246
(bottom), 247 (bottom), 249, and 268; National
Aeronautics and Space Administration, page
258; New Canaan YMCA, page 320;
Smithsonian Institution, pages 240 and 262;
Thiokol Chemical Corporation, page 264 (top).
All other photographs and drawings in this
book were supplied by the author.

ISBN 0-695-80615-7   Paper
ISBN 0-695-80616-5   Cloth

Library of Congress Catalog Card
Number: 75-13852

Fourth printing

Contents

Preface

Every author has a book or story that he has always wanted to write. In my case, this is it. I labor under the delusion that it might be more important than all of the fiction I have done and of broader consequence than much of my nonfiction.

I hope that it may save the hands, eyes, and lives of countless youngsters who might never have learned about model rocketry without it. I also hope that it may set many young people on their course toward becoming astronauts, engineers, technicians, and other kinds of scientists. Finally, I hope that it may serve as a guidepost to many people, young and old, who are interested in rockets.

Model rocketry is my hobby, to which I have given much time and effort and from which I have gained rewards far more valuable than mere money. The basic motive for my involvement in model rocketry stems from my youth, when many scientists, engineers, teachers, and other adults freely gave me advice, guidance, help, and the means to do things.

Once, I asked one of these men what I could do to repay him. "Do the same for others when you grow up," he told me. In the space-age hobby of model rocketry, I have found a way to do this.

G. HARRY STINE New Canaan, Connecticut August 1965

Introduction to the Fourth Edition

When this book was originally written in 1963 as the first comprehensive model rocketry manual, it was intended to serve as a complete handbook for model rocketry. I tried to cover every aspect of the growing young hobby to keep the beginner from "reinventing the wheel"—making the same mistakes as others had made and learned from. I also wanted to lead the modeler into the interesting advanced aspects of model rocketry.

The three previous editions, published in 1965, 1967, and 1970, respectively, have more than fulfilled these aims. The *Handbook* has had a major impact on the field. I know just how effective the impact has been from my conversations and correspondence with thousands of model rocketeers from all over the world.

As with any book based on any technology—and model rocketry is a technology in miniature—the *Handbook* has gone thoroughly out of date since the last edition appeared in 1970, just after men first walked the surface of another world, the moon. This current edition has required a complete, cover-to-cover rewrite and reillustration. And, because model rocketry has grown, the book can no longer cover *all* the aspects of model rocketry in great detail. The best I can do now is to present most of the basics and hope to lead you on to things beyond the scope of the book. The technical state of the art has advanced to the point where colleges and universities such as the Massachusetts Institute of Technology and the Air Force Academy accept undergraduate theses on model rocketry problems, as well as host annual technical conventions of model rocketeers.

This major revision is based on the fact that I have used the previous editions as texts for comprehensive model rocketry training courses given for eight years from 1965 to 1973 for the New Canaan (Connecticut) YMCA Space Pioneers Section of the National Association of Rocketry. This layout, format, approach, and content are based solidly on positive, in-the-field feedback from model rocketeers of all ages who had fun, who learned something, and who unknowingly contributed to this Fourth Edition in the process.

They have learned, as you and thousands of others have and will, that model rocketry is like wrestling with a bear: It is awfully hard to disengage yourself from it. No other hobby can claim to encompass as many different areas of modern science and technology. And no other aerospace hobby is as safe and as easy and inexpensive to participate in.

Model rocketry—or space modeling, as it is known outside the United States—has leaped the oceans to become a medium of communication between young people who, unlike their antitechnology contemporaries, got into model rocketry with starlight in their eyes. In the process they have "accidentally" discovered what science and technology are all about.

G. HARRY STINE Phoenix, Arizona December 1975

This Is Model Rocketry

Model rocketry has been called miniature astronautics, a technology in miniature, a hobby, a sport, a technological recreation, an educational tool—and it *is* all of these things. It is a safe, enjoyable, and highly respected pastime that now boasts the enthusiastic participation of millions of young people and adults in the United States, Canada, Great Britain, Australia, Sweden, Norway, Poland, Czechoslovakia, The Federal Republic of Germany, the German Democratic Republic, Rumania, Bulgaria, Yugoslavia, Egypt, and the Union of Soviet Socialist Republics.

Model rocketry was started in the United States in 1957, and its beginnings were carefully and completely documented. It resulted from a timely combination of model aeronautics, the ancient art of pyrotechnics, and modern space rocket technology. Although all of these elements had existed for over a decade prior to 1957, it fell to two men to combine them successfully into a space-age hobby.

The first model rockets were built and flown in 1954 by Orville H. Carlisle, the owner of a shoe store in Norfolk, Nebraska. Carlisle, with his brother's help, designed the first model rockets, putting together model aeronautics and ancient pyrotechnics. Early in 1957 I added the elements of modern space rocketry, and today's model rocketry was born.

Now, what is model rocketry, and what makes a model rocket so safe, so inexpensive, so easy to build, and so much fun that millions of people have successfully launched them? Why has model rocketry brought the space age to Main Street, directly involving more people in rocketry than have ever watched a space launch from the beaches of Cape Canaveral?

A model rocket is an aerospace model, a miniature version of a real space rocket, with *all* of the following characteristics:

1. It is made of paper, balsa wood, plastic, cardboard, and other nonmetallic materials without any metals as structural parts except where absolutely necessary.

2. It weighs less than 16 ounces and carries less than 4 ounces of rocket propellant, in accordance with federal regulations.

3. It uses factory-made, preloaded, nonmetallic, expendable solid propellant reaction motors that are replaced after each flight. This eliminates any hazard of mixing or handling dangerous rocket propellant chemicals. Or it uses a safe noncombustible cold propellant rocket motor fueled with Du Pont's Freon-12 and incorporates the proper pressure safety valves.

4. Its solid propellant rocket motor is ignited electrically from a distance of 10 feet or more using a battery and an electrical launch controller with built-in safety features. Cold propellant rockets can be launched by mechanical means.

5. It contains a recovery device to lower it gently and safely back to the ground so that it can be flown again and again by installing a new solid rocket motor or replacing the cold propellant.

The model rocket is that simple, yet it changed the nature of nonprofessional rocketry. Before 1957, when model rocketry was born in the dawn of Sputnik and the space age, nonprofessional rocketry was so dangerous that it was banned by law in many states. Today model rocketry is so safe that you can launch almost anywhere, often with no greater formality than simply checking with your local public safety officials so they know what you're doing. Because of its outstanding safety record, model rocketry enjoys the favor of the Federal Aviation Administration, the Food and Drug Administration, and the prestigious, safety-conscious National Fire Protection Association (NFPA). It also enjoys the enthusiastic support of the National Aeronautic and Space Administration (NASA), the United States Air Force, the United States Army, the Boy Scouts of America, the 4-H Clubs of America, the YMCA and the YMHA, and the Civil Air Patrol. Even the highly professional American Institute of Aeronautics and Astronautics (AIAA) reversed its early stand against nonprofessional rocketry to support model rocketry.

This confidence is deserved, for as of 1974, sixteen years after the first model rocket kits and motors became commercially available, over 50 million model rockets had been flown successfully and

Figure 1-1: Model rocketry is a national and international sport. Here a model rocket lifts off from the Mall in Washington, D.C., with the nation's Capitol in the background. The occasion was the annual Mall Demonstration of the National Association of Rocketry sponsored by the National Air and Space Museum of the Smithsonian Institution.

safely in the United States alone. The hobby is far safer than swimming, boating, baseball, football, and cycling. Since 1964 the members of the National Association of Rocketry (NAR) have been covered by a major liability insurance policy underwritten by a large insurance company; *not one claim* was filed in the first ten years of this coverage!

So don't let anybody tell you that model rocketry is hazardous to people or property. The facts prove otherwise. Naturally, it is possible to get hurt in model rocketry—if you are stupid, don't follow the NAR-HIAA Safety Codes, and don't read and follow instructions. Some people can get hurt while stamp collecting!

Model rocketry's excellent safety record is due primarily to the nonmetallic, lightweight construction of model rocket airframes and to the preloaded model rocket motor.

It is not obvious at first that you can learn anything about rockets if you use a ready-made motor. But the science of astronautics is more than a matter of rocket motors and propulsion. The business of making a model rocket motor is a very complicated, expensive, dangerous, and delicate affair that *must not* be attempted by anyone with less than several thousand dollars for equipment, an advanced college degree in chemistry, several years' experience in handling explosives, several acres of land as a safe place to work, and a very large life insurance policy.

(Three professional model rocket motor technicians have been killed while making model rocket motors. Thus, it's clear that the motor manufacturers take grave risks and assume the hazards so that model rocketeers can enjoy their hobby in safety.)

Model rocketry is like model aviation regarding motor construction. If you wanted to learn something about aeronautics by building a powered model airplane, it would be very silly and very expensive to build the entire airplane, including the gasoline motor. You have more fun and learn more by building only the model airplane and buying the motor ready-to-run. Then you know that you will have a motor that will work and, if you are a careful builder, an airplane that will usually get high enough to crash.

The same holds true for model rocketry. By properly using a model rocket motor and building a successful model rocket airframe around it, you will learn about thrust, duration, total impulse, specific impulse, grain configuration, thrust-time performance, and other rocket motor facts. In the air your model rocket is a free body

Figure 1-2: Model rocketry was started in the United States by the author (left) and Orville H. Carlisle (right), founders of the National Association of Rocketry, shown here comparing model rockets at the Ninth Annual Model Rocket Championships.

in space because you have launched it beyond the surface of the earth; its actions in flight are quite different than if it were on the ground on wheels or skids. This flight performance can tell you a great deal about bodies in motion. The fact that the model flies through the air will introduce you to the fascinating mysteries of aerodynamics, weight-and-balance, stability, and drag. In finding out more about how and why a model rocket flies as it does, and sometimes as it doesn't, you can start to delve into optics, structures, dynamics, electronics, meteorology, materials science, and many other technologies, plus the mathematical tools that make them useful to mankind.

A model rocket is wonderfully simple to build. You can use common hobby tools of the sort used to build model airplanes, model cars, and model boats. Even the materials employed in these other model hobbies are used in model rocketry. Many people think that a rocket has to be made out of steel or other metal, but that isn't so. Why make a rocket out of metal when you can make it cheaply out of paper and wood? Why spend a lot of money to buy a metal welding outfit when a twenty-five cent tube of model airplane glue will do the same job?

Besides ease of construction, there are other reasons for using nonmetallic materials. High strength and light weight have always been prime design goals for flying devices of any type. Paper and balsa wood meet these requirements well, and they can be used in a model rocket because a model rocket motor with its paper or plastic casing is barely warm to the touch immediately after operating. This is because paper is a very good insulating material and does not readily conduct heat. Although much lighter in weight than metal, paper and balsa are surprisingly strong. In fact, balsa has a higher ratio of strength to weight than carbon steel!

In addition, nonmetallic materials are much safer if something should go wrong during the flight of a model rocket. A model rocket made of paper, balsa, and plastic literally disintegrates if it happens to hit something. It absorbs the impact by destroying itself. Model rockets have been deliberately launched point-blank into sheets of window glass; these experiments completely destroyed the models, but didn't even scratch the glass.

All model rockets have recovery devices. Why bother about a recovery device? Because you want your model rocket to come down and land in a condition to fly again. After you have spent some money and taken some time to build a model rocket, you'll want to

Figure 1-3: This typical model rocket kit has all paper and plastic parts. It is a beginner's kit, the Estes Alpha-III with parachute recovery and molded plastic tail assembly. The finished model stands ready for flight.

get more than one flight out of it. Recovery devices are very easy to install, and they work well with exceedingly high reliability. Even a 1-ounce model rocket plummeting back to the ground from an altitude of 500 feet can hurt if it hits you. A recovery device eliminates this potential hazard. The mandatory use of recovery devices is one of the reasons model rocketry is so safe.

But why build model rockets so small? Why not build them big so that they are impressive, go very high, and carry lots of payload like the "big ones" at Cape Canaveral? Several reasons. Model rockets are built small for the sake of cost, simplicity, and safety. The price of a model rocket goes up even faster than a rocket as the size increases. Big birds get to be very expensive. As model rockets become bigger, they become more difficult to build. And as model rockets become bigger, they become less safe. Model rocket motors are limited in power for the sake of cost and safety. So the real challenge in model rocketry comes from discovering what you can do with a small amount of power. A model rocketeer doesn't use brute-force methods; he does careful work in design and construction to get the ultimate performance from a given amount of rocket power.

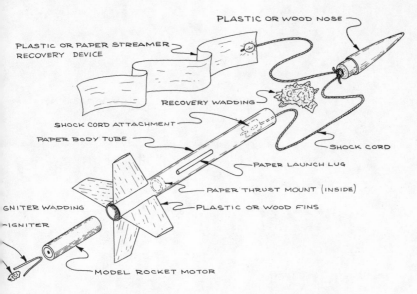

PLASTIC OR WOOD NOSE

PLASTIC OR PAPER STREAMER
RECOVERY DEVICE

RECOVERY WADDING

SHOCK CORD ATTACHMENT

PAPER BODY TUBE

SHOCK CORD

PAPER LAUNCH LUG

PAPER THRUST MOUNT (INSIDE)

IGNITER WADDING

PLASTIC OR WOOD FINS

IGNITER

MODEL ROCKET MOTOR

Figure 1-4: A sketch of a typical model rocket showing its various parts. There are many variations of the size and shape of many of the parts, but all model rockets are basically similar to this.

Models do not have to be big to carry interesting payloads. Using ingenuity and technical know-how, model rocketeers manage to fly and successfully recover fresh eggs, radio transmitters, still cameras, and movie cameras—all within the overall limitation of 1 pound of total weight at launching.

But why limit the rocket power? Doesn't this prevent a model rocketeer from launching models to really impressive high altitudes?

Yes, it does. But what are you going to do once the model rocket goes out of sight, which is something easily accomplished even with limited motor power? Small models go out of sight at altitudes between 1,000 feet and 1,500 feet; larger models are invisible above about 2,500 feet—less if the visibility is poor. Of course, it is possible to put a model rocket much higher than this. But why? Once it goes out of sight, you need radio tracking equipment, tracking telescopes, and large tracts of land. You might as well go to work for NASA in the first place . . . .

Sheer flat-out altitude is not the only goal in model rocketry. There are a lot of other things to do, things that are more fun, that are a challenge, that provide you with a real sense of accomplishment.

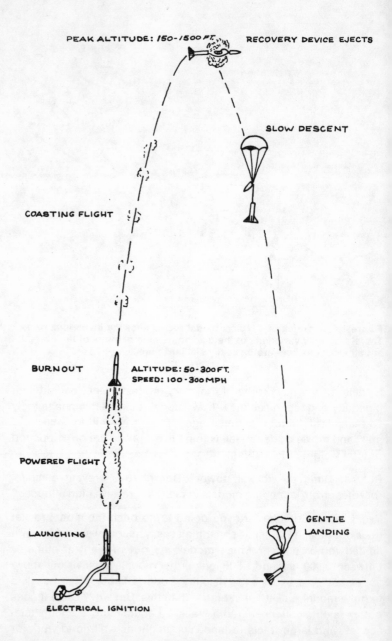

PEAK ALTITUDE: *150-1500 FT.*   RECOVERY DEVICE EJECTS

SLOW DESCENT

COASTING FLIGHT

BURNOUT   ALTITUDE: 50-300 FT.
SPEED: 100-300 MPH

POWERED FLIGHT

GENTLE LANDING

LAUNCHING

ELECTRICAL IGNITION

Figure 1-5: A typical flight of a single-staged model rocket with parachute recovery.

To begin with, you'll work hard getting your first model rocket together and off the ground for its first successful flight. You can do it—*if you will read and follow all instructions carefully.* If at first you don't succeed, try reading the instructions! After your first launch or two you'll be striving for reliable flights, reliable ignition, straight flights, full recovery deployment, recovery in the same county, getting it out of trees, keeping it out of trees, and getting it ready for more flights. You'll progress to more difficult models, to motors of higher power, to multistaged models, to glide-recovery models, and to payload-carrying models. You'll be able to try your hand at scale modeling. Or perhaps you'll design your own model and experience the thrill of seeing it work. Maybe you will enter contests and, after learning about model rocket competition techniques, start to win ribbons and prizes.

As you grow more deeply involved in model rocketry, you will find that you are studying a wide variety of subjects in order to understand what your models are doing and how to improve them. You will *never* have to worry about a science fair project if you are a student rocketeer.

When a model rocketeer talks with a professional rocketeer, they speak the same language. They can talk about specific impulse,

Figure 1-6: Recovery! At the end of its flight a model rocket lands so gently with its recovery device that it can actually be caught bare-handed in midair. Install a new motor and repack the recovery device, and the model is ready for another flight.

Figure 1-7: Model rocketry is international, having spread from the United States around the world. Here a Polish scale model of Poland's Meteor sounding rocket lifts off from the First World Championships for Space Models held in Vrsac, Yugoslavia, in 1972.

ballistics, launcher tip-off, data reduction, static testing, spin stability, recovery parachute deployment, lift-to-drag ratio, and hundreds of other topics. And, surprisingly, they have a great mutual respect for each other. In fact, many professional rocketeers are also model rocketeers. And many model rocketeers have gone on to become professionals.

This is because model rocketry is a technology in miniature, with many interesting aspects to consider other than the propulsion system. To model rocketeers, a model rocket motor is nothing more than a prime mover, a device that produces the thrust force to lift their models realistically into the sky, just like the "big ones" at the Cape.

Model rocketry is many things to many people. It can be a means to learn something about the universe in which we live. It can be a way to satisfy the competitive spirit that says, "My model is better than your model!" It can be a way to teach others. For many people it is an enjoyable recreation or hobby that combines the individualistic craftsmanship of the home workshop and the happy socializing with others on the flying field in the sunshine and fresh air. It can be a way for parent and child to get together in an activity of interest to both of them.

Model rocketry combines modern science and technology, craftsmanship and shop practice, individual creativity and group cooperation, and the pursuit of excellence with a healthful outdoor activity. Sportsmanship, craftsmanship, self-reliance, discipline, and pragmatic thinking are only a few of the things that can be learned in model rocketry.

But, mostly, model rocketry is fun. As long as you follow the safety codes.

And you'll never run out of things to do. Model rocketry is an endless countdown, a hobby that you can grow up with and stay with for years, for it offers a never-ending challenge to build a better model rocket.

Congratulations on getting hooked on the best aerospace hobby this side of Alpha Centauri.

Table 1
NAR-HIAA Model Rocket Safety Code

Solid Propellant

1. Construction—My model rockets will be made of lightweight materials such as paper, wood, plastic, and rubber without any metal as structural parts.

2. Engines—I will use only preloaded factory-made model rocket engines in the manner recommended by the manufacturer. I will not change in any way nor attempt to reload these engines.

3. Recovery—I will always use a recovery system in my model rockets that will return them safely to the ground so that they may be flown again.

4. Weight Limits—My model rocket will weigh no more than 453 grams (16 ounces) at lift off, and the engines will contain no more than 113 grams (4 ounces) of propellant.

5. Stability—I will check the stability of my model rockets before their first flight, except when launching models of already proven stability.

6. Launching System—The system I use to launch my model rockets must be remotely controlled and electrically operated, and will contain a switch that will return to "off" when released. I will remain at least 15 feet away from any rocket that is being launched.

7. Launch Safety—I will not let anyone approach a model rocket on a launcher until I have made sure that either the safety interlock key has been removed or the battery has been disconnected from my launcher.

8. Flying Conditions—I will not launch my model rockets in high winds, near buildings, power lines, tall trees, low-flying aircraft, or under any conditions that might be dangerous to people or property.

9. Launch Area—My model rockets will always be launched from a cleared area, free of any easy to burn materials, and I will use only nonflammable recovery wadding in my rockets.

10. Jet Deflector—My launcher will have a jet deflector device to prevent the engine exhaust from hitting the ground directly.

11. Launch Rod—To prevent accidental eye injury, I will always place the launcher so the end of the rod is above eye level, or cap the end of the rod with my hand when approaching it. I will never place my head or body over the launching rod. When my launcher is not in use, I will always store it so that the launch rod is *not* in an upright position.

12. Power Lines—I will never attempt to recover my model rocket from a power line or other dangerous place.

13. Launch Targets and Angle—I will not launch rockets so their flight path will carry them against targets on the ground, and will never use an explosive warhead nor payload that is intended to be flammable. My launching device will always be pointed within 30 degrees of vertical.

14. Prelaunch Test—When conducting research activities with unproven designs or methods, I will, when possible, determine their reliability through prelaunch tests. I will conduct launchings of unproven designs in complete isolation.

Cold Propellant

1. Engines—I will use only factory-made model rocket engines in the manner recommended by the manufacturer. I will reload rocket engines only with the propellant recommended by the manufacturer.

2. Recovery—I will always use a recovery system in my model rockets that will safely return them so they may be used again. I will conduct preflight tests to assure the recovery system functions properly before launching the rocket.

3. Weight Limits—My model rockets will weigh no more than 16 ounces at lift-off.

4. Stability—I will check the stability of my model rockets before their first flight except when launching models of proven design.

5. Flying Conditions—I will not launch my model rockets in high winds, near buildings, power lines, tall trees, low-flying aircraft, or under any conditions that may be dangerous to people or property. I will never attempt to recover a model rocket from a power line or other dangerous place.

6. Launch Rod—To prevent accidental eye injury, I will always place the launcher so the end of the rod is above eye level, or cap the end of the rod with my hand when approaching it. I will never place my head or body over the launching rod. When my launcher is not in use, I will always store it so that the launch rod is *not* in an upright position.

7. Launch Targets and Angle—I will not launch rockets so their flight path will carry them against targets on the ground, and will never use an explosive warhead nor a payload that is intended to be flammable. My launching device will always be pointed within 30 degrees of vertical.

8. Loaded Rockets—I will never store or leave a loaded rocket untended. I will always keep a loaded rocket on a launcher or firmly restrained. I will never point a loaded rocket or its rocket nozzle at anyone, nor allow anyone to be in the flight path of a rocket during flight preparations.

9. Construction—I will never use metal nose cones or metal fins.

Getting Started

You can get started in model rocketry for about ten dollars. This will set you up with a launch pad, an electric launch controller system, a launching battery, a simple beginner's model rocket kit, and a couple of model rocket motors.

All model rocket manufacturers (see Appendix I for their names and addresses as of this writing) make and sell starter sets that include all of the above equipment with the exception of the launching battery. You can purchase the equipment separately, but the pre-packaged starter set will save you time and money. In either case, you will continue to use the launch pad, electric launch controller, and launching battery in all your model rocket flying activities for a long time. So this big cost hits you only once.

If you are handy with tools, you can make your own launch pad and electric launch controller. You'll find the details of this in the chapter on launching.

Please don't try to fly a model rocket without a launch pad! And please don't attempt to use anything except electric ignition with a battery and launch controller in accordance with the instructions of the manufacturer. More than fifteen years' experience in model rocketry and over a quarter of a century of professional rocketry have proven the safety of using these two essential items. The few accidents that have occurred in model rocketry have involved attempts to shortcut the requirements for a launch pad and electric ignition. Friends, it isn't worth it. If you are going to do something, do it right or don't do it. And if you are stupid enough to think you are

21

smart enough to outguess all that technical experience and know-how, well, good luck and heaven help you!

You will probably need some help from your dad or other adult to assemble the launch pad and launch controller. However, the instructions that come with all starter sets are written for people who have never seen the equipment before, and most of it can be put together with such simple tools as a screwdriver and a pair of pliers. Remember: *Read the instructions first! Then build.*

While building, don't rush. Don't panic. Don't goof it. Do it right, and it will work right for you. I have helped thousands of model rocketeers get started, and I know that you are anxious to get that model into the air for the first time. But if you don't take the time to build your launch pad, electric controller, and model correctly according to the instructions, you are likely to be very disappointed when you finally push that launch button.

Current beginner's models are marvels of top-notch product engineering based on careful studies of beginner's requirements and on thorough field tests of the models in the hands of beginners. Just because the model is a simple tyro's bird, don't get the idea that it won't perform. Many beginner's rockets have contest-winning performance because of their simplicity. I have set national model rocket performance records with beginner's model kits.

Among the most ubiquitous beginner's models recently are the Estes Alpha-III and the Centuri Screaming Eagle.

From time to time you are likely to find ready-to-fly all-plastic model rockets in the stores. Some of these fly well, but usually good flight performance has been made secondary to appearance for sales appeal, ease of manufacture, or gimmickry. Such RTF model rockets are intended for people who just want something for kicks on a Sunday afternoon, not for someone like you who is interested enough in model rocketry to pick up this book and read this far in it! Sometimes all-plastic RTF models are good for demonstrations. But it's much more fun to build your own.

Although it often takes months to build a contest-winning scale model, you should be able to put together a beginner's model in 30 to 60 minutes. The speed record for assembling a beginner's model from the instant of opening the box to the moment of launch is held by Greg Scinto, of Stamford, Connecticut, who achieved the dubious honor of doing it in 11 minutes 56 seconds back in 1970!

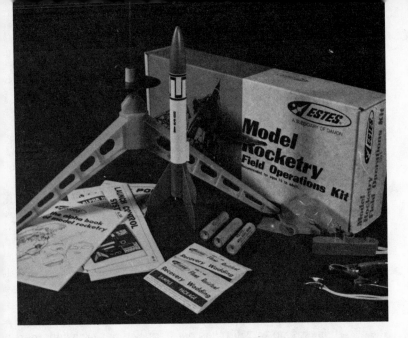

Figure 2-1: This typical starter set has nearly everything you will need to get started in model rocketry—a model rocket, a launch pad, an electrical controller, recovery wadding, three motors, and complete instructions. You need supply only simple tools, a battery, and the ability to follow instructions.

Most model rockets, regardless of their details of construction, size, and performance capabilities, have the following parts and assemblies (please refer to Figure 1-4):

1. A hollow plastic or balsa nose that fits onto the front of the hollow rocket body and will come off.

2. A hollow, lightweight, thin-walled plastic or rolled paper body tube that is the main structural part of the model rocket airframe. This body tube holds the recovery device, the motor, and the fins.

3. A launching lug, which is a small tube like a soda straw that is glued on the exterior of the body tube. It slips over the launching rod of the launch pad, thus holding the model on the pad before launch and guiding the model during the launch phase of the flight.

4. A recovery device, such as a plastic parachute or paper streamer, that is packed inside the body tube and is ejected forward from the body tube by a retro-thrust action of the model rocket motor at a predetermined time after launching.

5. The expendable, replaceable solid propellant model rocket motor and its associated thrust mount and retainer—or a cold propellant model rocket motor with its tank, nozzle, loading valves, and safety valve.

6. Fins made of balsa, plastic, cardboard, or pressboard that are fastened on the rear end of the model and, like feathers on an arrow, keep the model traveling in a true and predictable flight path.

7. Recovery wadding to protect the recovery device from the gas of the retro-thrust action of the motor and to help form a piston to eject the device from the front of the body tube. In cold propellant rockets the recovery wadding is replaced by an ejection spring and other devices to eject the recovery parachute at the proper time in the flight.

8. An electric igniter to start the solid propellant model rocket motor, or to remove the nozzle plug of a cold propellant model rocket motor.

The model rocket kit that you buy or that comes with your starter set will have all the necessary parts for you to assemble into a model rocket with all of these features. It will be up to *you* to assemble them into a strong, lightweight, streamlined model that will slip through the air at speeds up to 250 miles per hour without breaking apart because of air resistance. You will be assembling a model that flies faster than the fastest model airplane!

Don't let this shake you up. You can do it *if you read and follow all instructions!* Millions of other people have done it. And today's beginner's models are a snap. Just take your time and do it right the first time, for once that model rocket is in the air, you can't call it back to make a correction.

Figure 2-2: Two of the best beginner's models are the Centuri Screaming Eagle (left) and the Estes Alpha-III (right). Both have plastic noses and integral tail assemblies and use the same motor types.

Many tyros don't bother to paint their first model rocket. It isn't really necessary, but let me give you a tip: paint it anyway. Paint it a bright color such as vivid red, bright orange, or fluorescent orange. Reason: so that you can see it in the air *and* in the tree in which it will inevitably land. You will also be able to find it more easily on the ground should it happen to elude the clutches of the rocket-eating trees in the neighborhood.

Nearly all model rockets operate in the same basic manner, although performance can vary greatly due to differences in weight, size, shape, power of the rocket motor, and other factors that we will discuss later.

Finally, you are all ready for your first flight session! Excitement reigns supreme! But before you leave the house to go to the flying field, make certain you have all of the following items:

1. Launch pad.
2. Launch rod.
3. Electric launch controller.
4. The safety key for the launch controller. Put it on the end of a bright strip of cloth or ribbon so that you don't lose it.
5. Your model rocket. Some people have forgotten them.
6. Three to six model rocket motors of the proper size and power for your model.
7. An electric igniter for each model rocket motor, plus two or three spares J.I.C. (Just In Case).
8. A roll of paper tape such as masking tape.
9. A roll of cellophane tape.
10. Plenty of recovery wadding.
11. The ignition battery.
12. Your father.

Why the last item? Why your father or other adult? Simply to ensure that you are successful. Flying a model rocket is not difficult or complicated. But it requires that you do *everything* correctly in the proper sequence *before* you push the launch button. In that respect, it's just like a live big bird launch at the Cape. I highly recommend adult participation in the construction and flying of model rockets, because I know too well from long experience that a young rocketeer's natural enthusiasm and excitement over flying a model rocket often cause him to overlook some important point in the countdown sequence. Although many young people are perfectly competent to build and fly a model rocket all by themselves in complete safety, and although they may know much more about it than an adult supervisor, the double-check feature of adult participation can often prevent mistakes made in haste and excitement.

The professionals use a double-check system at the Cape for the big rockets. Why should model rockets be any different from the big ones just because of their size? Besides, it's kind of fun to know more about something than one's parents.

Your next task is to choose a flying field. It should be away from major highways and freeways, power lines, and tall buildings. Its size depends upon how high you expect your model to fly. For most beginner's models propelled by Type A, Type B, or Type C motors, a ground area about the size of a school athletic field is usually adequate on a calm day. On a windy day the model will drift farther in the wind and will need some downrange area to land in.

To calculate the size of the field that you should use, divide the expected altitude of the model by four. The launch field should have no ground dimension less than one-fourth the anticipated maximum altitude to be achieved.

Roughly speaking, a typical beginner's model with a Type A motor will go about 500 feet high; it will go about 900 feet with a Type B motor and about 1,400 feet with a Type C motor. This means that for flying Type C motors, you need a field with the shortest dimension of about 350 feet, which is a little longer than a 100-yard football field unless there are sizable end zones.

Set up your launch pad on the *upwind side* of the field so that the model will drift back across the field and land within its boundaries.

Important: After setting up your launch pad and electric ignition system, *test it.* Hook up one of the spare igniters that you brought. It should not glow red-hot until you have put the safety key into the launch controller and pushed the launch button. If your system can't pass this test, you've got troubles. You will have to troubleshoot the electrical system to find out what is wrong. If the igniter does not glow red-hot, it probably means that your battery is too weak or dead. But it could also mean that you have other troubles in the wiring. Read the chapter on ignition and launching for details about troubleshooting. If your ignition system passes this test, you can have confidence that the model will not take off while you are hooking it up—and that it will take off when you want it to!

Now prep, or prepare, your model. Insert the recovery wadding. Fold up the recovery streamer or parachute so that it slides *easily* into the body tube. Stuff the shock cord in on top. Put on the nose, making sure that you don't jam the shock cord between the body tube and the nose base. The nose should be able to slide off easily.

For your first flights use a low-power motor such as the Type A—a Type A3-2, A5-3, A8-3, or A8-4. Check the instructions of your model to make sure what type of motor to use; some larger beginner's models require a Type B motor for best flights. Nevertheless, use the lowest possible motor power. The first flight is a test flight. And you will want to get the model back so that you can fly it again. (Any fool can ram a Type C motor into his bird and lose it on the first flight; that is no outstanding achievement!)

Make certain that the motor cannot slide out of the rear end of the model. Most tyro models use a metal motor clip that holds the motor firmly in the body tube. If your model does not have a motor clip, wrap the motor with cellophane or masking tape until you have to force it into place. Be careful! Too much tape and too tight a fit may cause you to buckle the body tube when trying to get the motor in or out again later.

Install the igniter in the nozzle of the model rocket motor, following carefully the instructions that came with the motor package.

Now your model rocket is ready for its preflight safety inspection, given by the adult who is supervising the flight activities. Why bother? Simply because it will be too late to correct a mistake after you have launched the model. It will be on its way skyward, and you can't possibly catch it!

During preflight time always keep the safety key to the launch controller in your hand. This will ensure that nobody but yourself will be able to switch the electricity to the motor igniter. You will be sure that the circuit is safe as you hook up the igniter.

Slide the model down the launch rod; the launching lug should slip easily over the rod. The function of the launching pad should be obvious to you now. It supports the model during preflight operations and hookup. And it provides the initial guidance of the model after ignition while its airspeed is still too low for the fins to stabilize the model.

Clear the launch area. Hook up the igniter, clipping one of the micro-clips on the firing leads to each end of the igniter wire. Make sure you have a good connection. Make certain that the clips do not touch each other and that both of them are not shorting out through any metal launch pad parts. Get in the habit of keeping your fingers

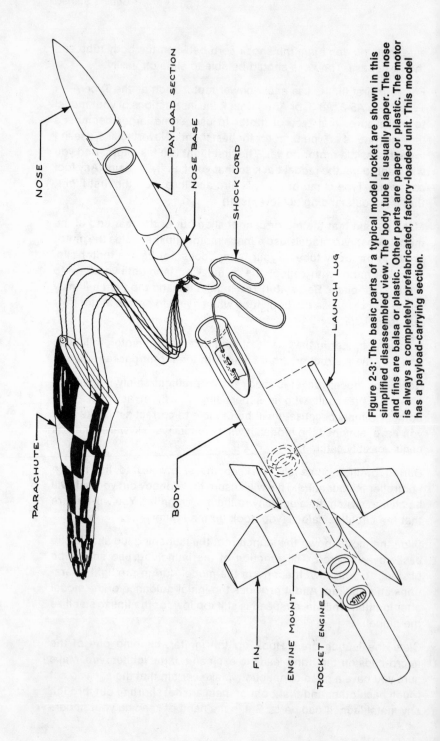

NOSE

PAYLOAD SECTION

NOSE BASE

SHOCK CORD

PARACHUTE

LAUNCH LUG

BODY

FIN

ENGINE MOUNT

ROCKET ENGINE

Figure 2-3: The basic parts of a typical model rocket are shown in this simplified disassembled view. The body tube is usually paper. The nose and fins are balsa or paper or plastic. Other parts are paper or plastic. The motor is always a completely prefabricated, factory-loaded unit. This model has a payload-carrying section.

Table 2
Altitude Range of Average Model Rockets

A small model rocket and a large model rocket were used to calculate the range of altitudes possible with a given motor type and a given lift-off weight. All altitudes are given in feet. Air resistance (drag) is taken into account assuming that a reasonably streamlined model rocket is used.

Motor	Lift-off Weight (ounces)	Altitude Range (feet)	Recommended Field Size (feet)
Type A	1	235 to 560	150 × 300
	2	170 to 260	150 × 300
	3	100 to 120	150 × 300
Type B	1	400 to 1,040	300 × 300
	2	370 to 760	200 × 300
	3	280 to 425	150 × 300
Type C	2	650 to 1,620	400 × 400
	3	600 to 1,270	350 × 350
	4	520 to 800	200 × 300
	5	460 to 580	150 × 300
	6	330 to 420	150 × 300

(A regulation football field in the United States is 50 yards wide and 100 yards long. This is a field 150 × 300 feet in size.)

out from under the direct line of the jet exhaust while doing this. Also, never look down on the model as it sits on the launch pad, hooked up and ready to fly.

Now retire to the launch controller. Check the area. If it is clear of people and all other interference, the adult safety officer should give you the "all clear" or "all systems go" for launch. Insert the safety key. The ignition continuity light in the launch controller should come on, telling you that you've got a hot circuit ready to launch. Give the old countdown: "Five . . . four . . . three . . . two . . . one . . . start!"

When you press the launch button on the controller, the electric current will flow from the battery to the igniter in the rocket motor. This will cause the igniter to glow red-hot instantly (or less than instantly if you have a weak battery). This, in turn, ignites the solid propellant in the model rocket motor.

The solid propellant begins to burn, creating large volumes of hot gas that rush out of the rocket nozzle, producing the thrust force that accelerates the model on its way.

The model will lift off very quickly. It will reach the end of the launch rod in about 2/10 second and be traveling about 30 miles per hour as it takes to the air.

Powered flight lasts for 1 second or less. During this time all the solid propellant in the motor is used up. At this burnout point in flight the model will be 50 to 200 feet in the air and traveling at a speed between 100 and 300 miles per hour straight up!

Now coasting flight begins. The model trades speed for altitude. The end of burning of the solid propellant has started a slow-burning time delay charge in the model rocket motor. This produces no thrust, but allows the model to coast upward.

At or near apogee, maximum distance from the earth, and at a predetermined time after ignition—chosen by the rocketeer who chooses the motor with the proper amount of time delay built into it—the recovery ejection charge in the motor activates. This produces a retro-thrust puff of gas, which pushes forward into the body tube and pressurizes the tube, forcing the wadding and recovery device forward and dislodging the nose. The recovery device deploys into the air and slows down the model. The model then returns slowly and gently to the ground. Since all parts are tied together by the shock cord, the entire model lands together where it can be easily recovered—unless it has landed in the one and only tree in sight, naturally.

If the model lands in a power line, forget it. Don't ever try to get a model rocket out of a power line. You can buy a new model for a couple of bucks. You cannot buy back your life.

With your model happily in hand, return to the launch area. Check the model for damage such as broken fins or cracked parts. Take out the expended model rocket motor casing and put it in your field box; take it home and throw it away later. Don't give it to the little urchins who will immediately come out of the bushes within minutes after

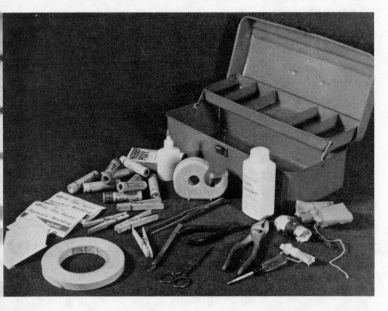

Figure 2-4: In addition to model, launch pad, controller, and battery, the fully prepared model rocketeer always carries a complete range kit with all the little tools, pieces of equipment, and other goodies that make flying easier and more fun.

the first launch; they may try to stuff it full of match heads and get hurt. Repack the recovery device. Install a new model rocket motor and igniter. Present the model for its safety check. And you're ready to put it in the air again!

I have seen model rockets that have made over one hundred flights—but not in one afternoon! You will be doing very well to get off three or four flights in your first afternoon. Try higher-powered motors and see the difference. Experiment with tilting the launch rod a few degrees away from the vertical.

Once you have put in your first afternoon of flying model rockets, you're ready to go on to bigger and better models. I am certain that you will be hooked at this point. Your dad may discover that he's hooked, too! I have known many fathers who grumpily went to the flying field for the first flights because their young rocketeers insisted on doing things the right way—and I have seen them come home afterward more excited about building and flying model rockets than their offspring! Model rocketry is not just a kid's hobby!

A model rocketeer's first flight is an important one, and he or she remembers it even after flying thousands of models over the years. I can still remember my first flight one chilly February day in a cotton field in Las Cruces, New Mexico, in 1957. Today, many thousands of flights later, I still get the same kick out of pushing that launch button and watching one of my creations lift off into the sky.

Sure, it won't get to the moon—but I can imagine it does!

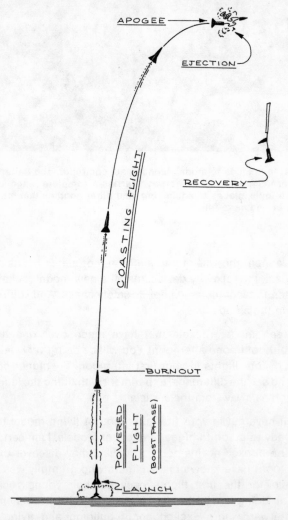

Figure 2-5: This sketch shows the basic flight phases of a model rocket. This is the sort of performance you should expect of your first model rocket—and every one you build thereafter, too.

Figure 2-6: Lift-off! The moment of truth arrives as you push the launch button, ignition occurs, and your model rocket starts to streak aloft from its launch pad.

Tools and Techniques in the Workshop

In working with model rockets, common sense is required. If you don't have it to start with, you develop it very quickly! You can always develop your manual dexterity and ability to work with tools, also.

The most important rule for becoming a successful model rocketeer is: *Be willing to read and follow all instructions completely!*

I cannot emphasize this too strongly. The biggest mistake made by novice model rocketeers (and even some expert model rocketeers who think they know better but don't) is failing to read and follow instructions. As an indication of what can happen when a rocketeer gets rushed or slapdash in his work, or becomes a victim of "I-know-it-all," I have actually seen the following gross mistakes in beginners' model rockets:

1. A nose glued to the body tube so that the nose could never come off for the recovery device to deploy.

2. A motor mount glued in backwards so that it was impossible to insert a model rocket motor of any type into the model.

3. A solid bulkhead glued across the inside of the body tube so that no ejection charge gas could eject the recovery device.

4. Recovery wadding glued into the body tube.

5. Balsa fins that were cut out with a pair of scissors. "But, Mr. Stine, it makes such a nice, fuzzy edge for gluing!"

6. A model rocket motor firmly glued into the model.

7. Finally, the endless, disheartening parade of model rockets with crooked fins, fins that are too small, fins glued on the nose of the model (a definite no-no), no launch lug, launch lug mashed flat, no motor mount, shock cord not glued in, and (yes!) body tube bent.

Such models are garbage birds good only for recycling or to go into the trash can. I know that *you* can do better!

Once you have built and flown your first beginner's model, you should go on to more difficult kits and, eventually, to designing your own. First of all, though, give up every notion you may have had about rockets. Take that design for the four-staged radio-carrying mile-high rocket and tuck it away in your notebook; you aren't ready to build it yet. You have a lot to learn, and you will have a lot of fun learning it. Later, when you look with greater knowledge at that "early, primitive" design, you will either laugh yourself silly or discover that you need to make a lot of changes based on your experience. This is true even if you are an experienced professional rocketeer working for NASA or USAF who has taken up model rocketry as a hobby or if you are a championship airplane modeler who has decided to try something new.

We do many things differently in model rocketry, and we do them for a purpose. You will learn *why* in this book. You will also be able to see why in your own workshop and out on the flying field.

Workshop

First, you must have a place to work. Beginner's models and some other simple kits don't require much of a workshop. But if you want to progress beyond simple models, you will have to develop a workshop. It is a place where you can keep your tools, your model rocketry supplies (spare noses, body tubes, balsa, and the like), your models as you are building them, and your models once they are finished.

A card table in your bedroom is all that you may need to start with. But balsa dust, wood chips, paint overspray, and other debris accumulates while building model rockets, and you really need a place that will keep this stuff from spreading itself thinly all over the house to the dismay of mother or wife. In addition, modeling paints and glues tend to be somewhat aromatic. In other words, to a person who isn't a modeler, they stink.

If you can manage to commandeer a corner of the basement as a workshop, be sure that there are windows that you can open to

Figure 3-1: A well-lit, well-equipped workshop is essential for good model rocketeering.

provide ventilation. The days of glue sniffing are over now because the manufacturers have included in the glue a substance that makes you violently sick to your stomach if you inhale too much glue vapor. The vapors of drying dope and enamel paints aren't good for you, either. So make sure you have adequate ventilation.

Garages make excellent workshops in warm climates where you don't freeze in the wintertime. If you live in a colder area and can get a garage as a workshop, it should be heated, for it is very difficult to do precision model work with frostbitten fingers or with gloves and other cold-weather clothing on. But keep paints, glues, solid propellant model rocket motors, and cold propellant cans away from the heater unless you want to put the garage into orbit.

Some of us, myself included, are lucky enough to have a large workshop because the woman of the house is convinced that model rocketry is a great father-son activity that keeps the men of the house happy and out from under foot. And some households boast having the whole family involved in model rocketry, which is an ideal arrangement for getting the best possible workshop area in the house.

Wherever you manage to set up a workshop, try to keep it in some semblance of order and cleanliness, impossible as that may be at times. This keeps you from tracking balsa dust into the house. It also permits you to find things more quickly when you are in that panic-rush to complete a model the night before a major contest— and to find that little bitty part that follows Paul Harvey's Law: A dropped part will always roll into the most inaccessible part of the workshop. It also greatly impresses your friends. Mostly, though, a neat workshop makes it easier for you to get along with the other nonmodeling members of your family.

Simple tools

Your most powerful tool is a notebook in which you can write down your ideas, make sketches, note how you did things so that you can remember them better, and file technical reports and other papers. I have twelve loose-leaf notebooks, each of them full of information.

Figure 3-2: You can set up a simple workshop on a card table in your bedroom, but be sure to use newspaper to protect furniture and refrain from making lots of balsa dust.

They are the most valuable books in my library. *The Handbook of Model Rocketry* was compiled from information that I had tucked away in them. Scientists and engineers always keep notebooks and logbooks for their ideas, progress notes, reports, and other data. Model rocketry is a scientific hobby, so it will pay you to start keeping a model rocketry notebook from the very beginning. There is a tremendous amount of information available in model rocketry. You don't want to lose it, so put it away where you can find it again when you need it—in your notebook. Your notebook will keep you from "reinventing the wheel," which is the mistake of duplicating somebody else's mistake that you should have read about and noted.

Since you cannot put a model rocket together with your bare hands, you will need a few simple tools, and you will have to learn how to use them. You may already have some model rocketry tools. All of them can be purchased quite inexpensively at your hobby or hardware store.

The use of tools is one of the many things that sets human beings above the beasts of the jungle. So get good tools to start with, treat them properly, and use them safely.

You will need a *modeling knife.* This can be a new single-edged razor blade; don't use a double-edged razor blade because you can easily cut yourself with it. You can buy a modeling knife at the hobby store for a dollar or so, and it will last you for years. It has a slim metal handle with a holder, or chuck, at one end to grip a special extra-sharp blade that will cut plastic, paper, and balsa. A typical modeling knife is called an X-Acto knife. One of the best knives for rocket modeling is the X-Acto No. 1 knife. I like to use it with the X-Acto No. 11 blade. Buy a package of extra blades for your knife so that you will always have a sharp blade. When the blade gets dull, take it out and install a new blade.

Note: When you use a modeling knife, be careful. It cuts fingers more easily than it cuts balsa.

A *pair of tweezers* is very handy for holding parts that are too small for your fingers to hold; thus, they become a very fine extension of your fingertips. You can also use them to reach into small places to grasp. Tweezers can be purchased in a hobby store. You may also wish to get one or two pairs of tweezers that unclamp as you squeeze them so that they can be used as long, thin clamps.

A sharp *pencil* or a *ball-point pen* is a must for making notes, marking parts, etc. To keep it from "walking away" from your workbench, put it on a leash; tie or glue one end of a 24-inch length of string around it, and attach the other end to your workbench with a tack, nail, or staple.

Scissors are useful for cutting out paper templates, decals, and other paper parts. As mentioned before, please don't use them to cut balsa! You may also have to attach your scissors to the workbench so that they don't get appropriated by someone else.

Needle-nosed pliers are useful for assembling things with nuts and bolts, for holding parts, and for bending metal. Again, your hobby store is a source for pliers.

A *small screwdriver* is a model rocketeer's friend in the shop and in the flying kit that you take to the flying field.

A metal 12-inch or 6-inch *rule* will be used to measure and to guide you in cutting straight lines with a modeling knife. Don't get a plastic or wood rule; you will end up cutting its edge and will therefore make balsa cuts that look like a plan of a mountain highway. To cut a straight line with a modeling knife, merely lay the steel rule down on the balsa, cardboard, or paper, and run the knife right along the rule edge.

If you can find a metal rule with both English system inches and metric system millimeters or centimeters on it, you will discover how easy it is to measure things in the metric system. A millimeter, for example, is 40/1000 of an inch (0.040 inch), or roughly 25/64 inch. It is much easier to work in the round numbers of the metric system than in the fractions of an inch of the English system. For example, it

Figure 3-3: These are some of the simple, everyday tools that you will use most in building all sorts of model rockets.

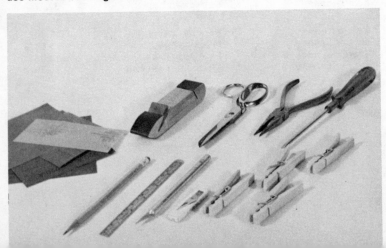

Table 3
Metric Conversion Chart

From	To	Multiply By
Centimeters	Inches	0.3937
Centimeters	Feet	0.0328
Inches	Centimeters	2.54
Inches	Millimeters	25.4
Inches	Meters	0.0254
Meters	Inches	39.370
Meters	Feet	3.2808
Meters	Yards	1.0936
Feet	Centimeters	30.48
Feet	Meters	0.3048
Feet	Kilometers	0.0003048
Grams	Ounces	0.03527
Grams	Pounds	0.0022046
Kilograms	Ounces	35.2739
Kilograms	Pounds	2.2046
Ounces	Pounds	0.0625
Ounces	Grams	28.349
Ounces	Kilograms	0.028349
Pounds	Grams	453.592
Pounds	Kilograms	0.45359
Newtons	Pounds-force	4.45
Pounds-force	Newtons	0.2247
Pounds-weight	Pounds-mass	0.03105
Pounds-mass	Pounds-force	32.27
Miles per hour	Feet per second	1.467
Feet per second	Miles per hour	0.6818
Meters per second	Feet per second	3.281
Feet per second	Meters per second	0.3048

is much easier to work with 15 millimeters than with its close equivalent, 19/32 inch! Model rocketry is technically on the international metric system and has been since 1962. However, many measurements are still given in the English system of inches, pounds, etc., because most Americans are familiar with these units. In model rocketry you can become quickly and easily acquainted with the metric system, and so will be prepared when the United States converts to it before 1985. Take this opportunity to learn metrics.

A good stock of *sandpaper* of various grits should be kept on hand. It is used for shaping and smoothing. It comes in various grades ranging from very coarse to very fine. For model rocket work you can buy large sheets of sandpaper at the hardware store at very low cost. Cut these big sheets into little sheets about 2 inches (50 millimeters) square, using a pair of scissors. Get sheets of No. 200, No. 320, and No. 400 wet or dry sandpaper. The No. 200 is useful for shaping, the No. 320 is good all-around stuff, and the No. 400 is great for final smoothing.

You can buy a *sandpaper block* at the hobby store, or you can make one by taking a convenient-sized wood block and tacking some No. 200 sandpaper around it. The wood block provides a flat, firm base for the sandpaper and permits you to shape flats, curves, and weird shapes more easily.

You will need some *clamps* to hold parts together while they are drying, to hold parts generally, and to use around your launch pad on the flying field. The finest modeling clamp known to man is the good old-fashioned spring clothespin. Before most people had

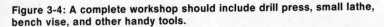

Figure 3-4: A complete workshop should include drill press, small lathe, bench vise, and other handy tools.

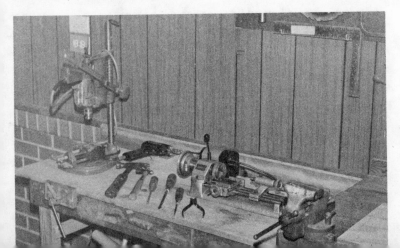

clothes dryers, these clothespins were common household items. Today you may have to buy some in the hardware store. Once you have them, you will find all sorts of uses for them because, being wood, they can be sawed, drilled, carved, sanded, painted, glued, and made into many needed items.

These are the basic tools. I have built a lot of very good model rockets in a one-room apartment or in a motel room with no tools other than these.

The ideal workshop

Every model rocketeer dreams of the ideal workshop. I have one, but it took years—plus some cash—to accumulate all of the tools, jigs, fixtures, and gadgets.

Probably the most ubiquitous power tool in American homes is the 1/4-inch or 3/8-inch *electric drill*. This is a very handy model rocket tool if you use some ingenuity to set it up in various ways. Sure, it will always drill holes if you put a drill bit in the chuck. But by using a drill press accessory available where you bought the drill, you can turn a simple electric drill into a very effective and accurate drill press, buffer, polisher, and grinder. With a horizontal holder, the electric drill becomes a grinder or a useful miniature lathe for turning model rocket parts out of balsa. Many modern electric drills have variable speed controls and reverse features; they shouldn't be called just "electric drills" because they are, in reality, portable rotary electric power sources. Think of your electric drill in those terms, and you can find lots of uses for it.

If you have a small wood lathe or can get access to one in the school shop, you are really in business! Such a lathe can be used to turn many model rocket parts. Although the Austrian Unimat lathe is fairly expensive, I have used one since 1957 in my model rocket building, and it has paid for itself many times over because I have been able to make a lot of parts instead of buying them. There are other small, inexpensive wood lathes, too. Of all the power tools in a model rocket workshop, the lathe is perhaps the most used and most useful.

Other handy tools include a *claw hammer, a small ball peen hammer, standard slot and Phillips-head screwdrivers* of various sizes, *side cutters, diagonal cutters, soldering iron,* model maker's *tee-head pins,* a *vise,* and a *building board.*

A building board is a rectangular piece of plywood, composition board, Homosote board, cork, drywall, or the cutout from a Formica countertop available at some lumberyards. The board provides a flat, smooth surface to work on, one that you can cut on without scarring the tabletop and some of them you can easily stick pins into.

All sorts of little jigs and fixtures have been developed by model rocketeers over the years. Here are some of the ones that I have found to be most helpful.

The cradle stand: How do you set a model rocket down on the workbench while you're working on it? Answer: Do as the NASA sounding rocket people do and use a cradle. A typical cradle that can be cut from a piece of balsa is shown in Figure 3-5. I started using cradles like this in 1957, copying them from the ones we were using on the full-sized Aerobee sounding rockets at White Sands, where I was then working. If you have a model that is shorter or longer than the cradle shown in Figure 3-5, you can make a shorter or longer cradle. I have a series of cradles ranging from 3 inches to 9

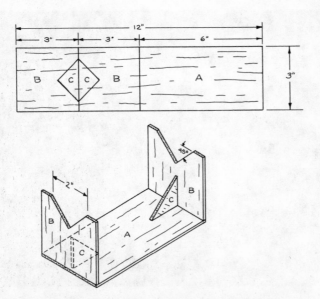

Figure 3-5: A handy fixture is the cradle stand that can be cut from a piece of sheet balsa as shown. It will support models horizontally.

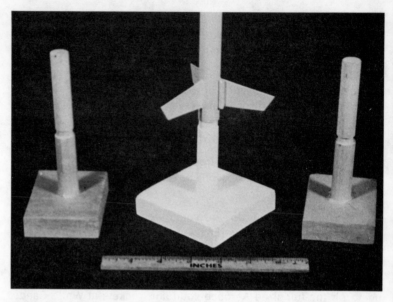

Figure 3-6: The workshop spike can be made from wood and is designed to hold the model vertically.

Figure 3-7: A spike row is a line of spike dowels on a single board for holding several models vertically or for storage or carrying.

inches long. Sometimes I paint them if they are to be used to display a model. You will find that a cradle is one of the handiest fixtures in the shop.

The spike: Wendell H. Stickney built the first spike in 1961 to hold a model rocket in an upright position during construction or for display. The easiest way to make a spike is to glue an empty motor casing to a block of wood. However, you can make a better one on a lathe. A typical spike is shown in Figure 3-6. It is longer than a regular motor casing and will hold the model high enough to clear any swept-back fins. A group of spikes will increase your workbench space because they hold the models vertically.

Spike row: The obvious and logical extension of the spike is several spikes fastened on a long piece of board. This is useful not only in the shop or for an impressive display, but is handy when you have to carry a lot of models. It is as easy to make as a single spike. With the advent of the mini-motors, I have made all my new spikes and spike rows with 1/2-inch wood dowels; they fit nicely inside the mini-motor mounts of mini-models. To use them with standard motors, slip an empty thin-walled 19-mm standard motor casing down over the 1/2-inch spike. Or use the motor spacing tube that comes with some Estes or Centuri kits.

The paper wand: How do you spray paint a model rocket without getting the overspray all over your hands? Make a paper wand by rolling up a sheet of newspaper and fitting one end of it into the rear end of the model. Hold onto the other end. Spray only at the model end, of course! The flimsy newspaper page is quite strong after you have rolled it into a wand. Figure 3-8 shows a model being spray painted using a paper wand.

The rotating spike: For painting models with an airbrush, spray gun, or spray can, the Stines built a spike mounted horizontally on a support that would also permit the spike to rotate. The spike is inserted into the model's motor mount. The model is thus held horizontally and can be rotated like a roasting pig on a spit over a fire. This permits easy spray painting of the entire model. The gadget requires some work with a lathe and a drill press, but it is not so complicated that you can't make it in a couple of hours as a quickie project.

Fin alignment angle: How do you draw a line accurately along a model rocket body tube so that you can correctly locate the fins? The answer is so simple that you may not believe it. Go to a hobby or

Figure 3-8: A paper wand made from a sheet of rolled-up newspaper stuck in the rear end of the body tube is handy when spray painting.

hardware store and buy a small piece of metal angle about 6 inches long. You should get 1/2-inch angle or smaller. If you can find some old brass angle used to make slot car bodies, it's perfect. Simply lay the angle down along a body tube as shown in Figure 3-10. Use one edge of the angle as a ruler to draw a straight line perfectly aligned

Figure 3-9: A rotating spike holds a model horizontally so that it can be turned to spray paint all sides of it.

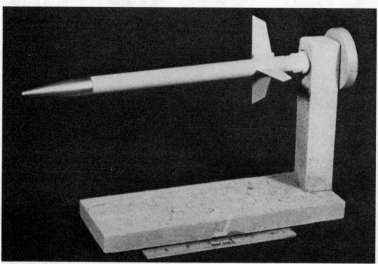

with the long axis of the tube. Both the angle and the line *have* to be aligned with the body tube!

Fin positioner: How do you get three, four, or six fins properly located around a body tube? Simply make a fin positioner similar to the one shown in Figure 3-11. Center the body tube on the positioner as shown. Mark the proper location of the fins on the rear end of the body tube. Then use your fin alignment angle to draw the positioning lines on the tube.

Fin jig: In the history of model rocketry thousands of different fin assembly jigs must have been designed to help the beginner get his fins on straight. The best one that I have ever seen was invented by Howard R. Kuhn, many times United States Senior Champion and a member of the United States Space Model Team for international competition. You can buy one of Kuhn's jigs from Competition Model Rockets at the address listed in Appendix I. Or you can make one. Go to your hardware store or lumberyard; get a piece of wood angle as shown in Figure 3-12. Cut off about 6 inches. With a saw cut a slot at the apex of the angle, as indicated. Make the slot 1/16-inch wide for normal fins; you can make another jig with a slot 3/32-inch wide for heavier fins. Put the angle on the body tube as shown, and hold it in place with two rubber bands around the model. Insert the fin through the slot as shown, and the jig will hold the fin in perfect

Figure 3-10: A small brass or steel angle laid against the side of a body tube permits you to draw a straight line down the tube for positioning fins.

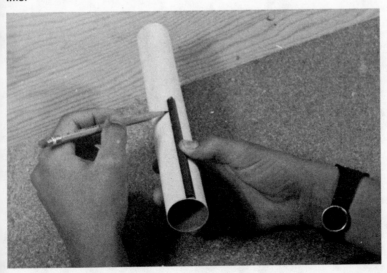

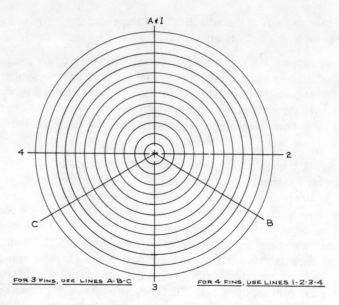

FOR 3 FINS, USE LINES A-B-C   FOR 4 FINS, USE LINES 1-2-3-4

Figure 3-11: With a fin positioning drawing such as this, fins can be properly located on the tail of a body tube.

Figure 3-12: The Kuhn fin jig can be made from a piece of wood angle and will hold a fin straight until the glue dries.

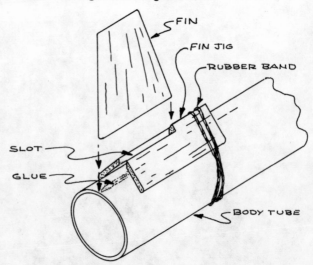

position until the glue dries. The design of Kuhn's fin jig eliminates the worry about gluing the jig to the model.

Q-Tips: Often you will need to put a bead of glue around the inside of the body tube in order to glue a motor mount in place. The easiest way to do this is with a tool swiped from the bathroom medicine cabinet and widely known by the trade name Q-Tip. It is a wood or paper stick with a ball of cotton on one or both ends. You can easily make one using a dowel and cotton. Or you can buy a box of them at the drugstore. Mark on the stick the distance inside the body tube that you want the glue to be located. Put glue on the cotton swab and carefully insert the Q-Tip down into the body tube, being careful not to get glue on the inside walls until the Q-Tip has reached the proper depth. Then swab the glue around the inside of the tube. It may take several applications before you get a good band of glue in there. But you can use the same Q-Tip many times. In fact, it gets better as you use it because a hard glob of dried glue builds up on the cotton ball, making it easier to use.

There are many other special tools, jigs, and fixtures that have been developed by other model rocketeers. I will probably hear very shortly about all of those I haven't mentioned here! But I have told you about those items that I have tested and found to be most suitable and useful to most model rocketeers.

Glues

There used to be only one kind of glue to use in building model rockets—model airplane glue. But the field of bonding technology has so progressed that many types of adhesives are available today. We still call them "glues" although in a true technical sense many are "bonding agents." They give the model rocketeer a wide choice to use with many different types of materials.

The materials used in model rocketry are not difficult to join by bonding or gluing. But you must use the proper bonding agent. Porous materials such as paper, cardboard, and balsa require a different type of bonding agent than nonporous materials such as plastic and metal.

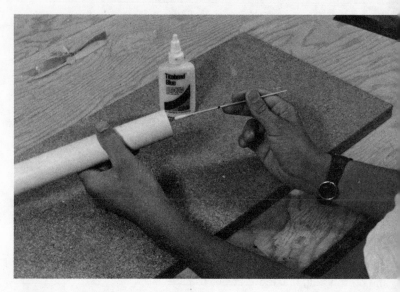

Figure 3-13: A cotton swab, or Q-Tip, can be used to apply glue to the inside of a body tube. Mark the swab stick to tell you how far into the tube to apply the glue.

Figure 3-14: A wide variety of glues can be used in constructing model rockets, but some are intended only for special applications.

The following information on glues and bonding agents has been accumulated by years of model building and by correspondence with aerospace modelers from all over the world.

Model airplane glues such as Testor's Formula B, Pactra C-77, Ambroid, and Du Pont Duco Cement are used for gluing balsa to itself or to fiber or paper body tubes. They will also glue paper together. They will glue some plastics but are not considered to be optimum for use with plastics.

White glues such as Elmer's Glue-All are casein-based glues made from milk products that are water-soluble. They are exceptionally strong for use with paper, wood, and other porous materials. But they will not work on plastics. They take a long time to set up firmly and even longer to dry completely—overnight in most cases. You may have trouble applying some types of paint over white glue layers.

Aliphatic resin glues such as Franklin Titebond and Centuri Superbond look like brownish-yellow versions of white glues. But they have a different chemistry. They are very strong and will set up faster than white glues. They are transparent when dry, and most types of paint will go over them nicely. Aliphatic glues are very useful on porous materials but will not bond plastics.

Contact cements are reasonably new to the hobby field. They include rubber cement, Goodyear Pliobond, and Weldwood. To use them, apply to both surfaces, allow to dry for a few minutes, then join the surfaces. The surfaces will bond firmly *immediately* upon contact, so be careful. You haven't any leeway on moving things around to line them up because the contact cements grab and hold on contact. They will usually bond anything to anything else, but it is always best to test first on a couple pieces of scrap.

Epoxy bonding agents have revolutionized the model airplane construction field, but are just beginning to find their place in model rocketry. Basically, an epoxy is a type of plastic, technically known as a thermosetting plastic. Epoxies come in two parts—a resin and a hardener. When you are ready to use the epoxy cement, mix together the proper amount of each part to give you just the amount of bonding agent you need right away. Be sure to follow the instructions that come with the epoxy cement you buy. Don't mix too much epoxy because it cures, or hardens, in minutes. There are some 60-second epoxies, but the most commonly used types cure in 5 to

10 minutes. And when they harden, they are hard! Epoxy bonding agents are incredibly strong. Because they will bond nearly anything to anything else, they are very useful when you have to join a metal piece to a wood or plastic piece.

Cements or glues for plastics work in a different manner than glues for porous materials. *Plastic cements* actually melt or soften the plastic material so that when two plastic parts are pressed together, the plastic flows together and welds, becoming one piece of plastic. Several chemicals work well as cements for plastics, and most of them are solvents for plastic. They come in liquid or tube form.

Professional plastic model builders buy their plastic cements in large cans of a gallon or more. They use acetone or methyl-ethyl-ketone (MEK). You can go this route if you have 15,674 plastic models to assemble; otherwise, buy the liquid plastic cements in the small bottles available in the hobby store, and keep the caps on tight when you aren't using them. They are all highly volatile and will evaporate quickly if left open.

Plastic cements also come in tubes like some types of wood cements. This tube-type plastic cement is actually plastic with lots of solvent in it. When the solvent evaporates away, the cement becomes solid plastic. When you apply it to a plastic piece that will be joined to another plastic piece, the glue forms a bridge of plastic to unite the two parts. Testor's Cement for Plastic Models is available in hobby stores in an orange-and-white tube. It is intended for use with the polystyrene plastics that most plastic model kits are made of, and it really doesn't bond well to anything but polystyrene and its immediate derivatives such as cycolac. One of the very best plastic cements is Du Pont No. 9011 Plastic Cement, which comes in a blue-and-white tube and is available in hardware stores. It works on nearly all types of plastics except the waxy polyethylene and Teflon, but these plastics are not often used in models. Du Pont No. 9011 is the best glue to use for gluing transparent acrylic fins on model rockets. Both Testor's and Du Pont work equally well on wood and paper, although they take longer to dry than wood-type glues. These tube-type plastic cements should be used when you have to bond a paper or wood part to a plastic part because they penetrate the porous surface of the paper and permit a strong bond when they dry.

If you do not know whether the glue you have will work with certain types of materials, make some tests using scraps of the materials. Glue them together and see what happens. This doesn't take very long to do, and it may save you all sorts of grief—such as ruining your model in the construction phase or having it come apart in flight at 300 miles per hour.

There is a humorous side to bonding techniques, too. In aviation there is the speed of sound, and in physics, the speed of light. In model rocketry there is a similar barrier speed known as the speed of balsa. This is the speed at which balsa construction fails in flight. It is a variable constant because it depends totally on the individual model builder. Some people continually build models that exhibit a very low value of the speed of balsa.

You can increase the speed of balsa enormously by learning how to make a strong glue joint. Having a good glue is only half the battle; knowing how to use it properly is the other half. Almost 90% of the people who make model rockets do not know how to make a good glue joint, even though it is usually printed in the kit instructions. This is probably because they think that the correct method is nothing more than a time-honored ruse to get them to use, and therefore to buy, more glue.

If you want to make a glue joint that is stronger than paper or balsa, follow these simple instructions:

When gluing together porous materials such as paper or wood, use a double-glue joint. Coat both surfaces with a layer of glue and let it dry until it is firm to the touch. Then coat both surfaces *again* and join them together. The first coat of glue on both surfaces penetrates the pores of the materials. The second coat of glue is then free to join with the first coat and with the second coat on the other surface. A double-glue joint will be so strong that the materials will break or tear before the glue joint breaks. Try it. It will surprise you. And you will never use any other type of glue joint again.

When gluing together two pieces of plastic or other nonporous material, use a variation of the double-glue joint. Coat both surfaces with glue. Since the materials do not have pores, there is no need to apply two separate coats of glue to each surface. The cement will soften each surface so that they will weld together easily when joined. Again, such a joint is as strong as the material itself because that is what it is!

Having now become an expert in workshops, tools, and construction materials, you can tackle the building of model rockets with a great deal of confidence that *perhaps* everything will work right and nothing will go wrong—if you practice what you have learned.

Model Rocket Construction

Today most model rockets are built from kits that contain all the parts and components necessary to complete the model. Many of these parts are prefabricated, and the model rocketeer isn't required to complete any highly detailed parts that call for special techniques or power tools such as a lathe. However, there remains a small cadre of dedicated—some would say "crazy"—model rocketeers who prefer to build from scratch. They will sometimes use parts available from the manufacturers, but otherwise prefer to make everything themselves—except for the model rocket motors; making those oneself is not being purist but suicidal.

Figure 4-1: Assembling your own model rocket either from a kit or from basic parts to your own design requires some craftsmanship that you will improve upon as you build more models.

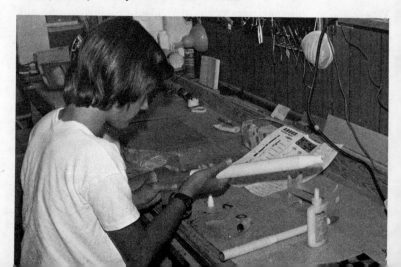

However, it doesn't make any difference whether you are a maker of prefab kits or a scratch-builder; there are certain techniques that are universal and used by both kinds of modelers.

I am devoting an entire chapter to this general subject because, in my large collection of hobby how-to books, most authors seem to assume that their readers are experienced and don't need basic explanations of how and why. I find the *why* to be very important to the technically curious type of person who becomes involved in model rocketry.

So, as our British colleagues say, "Back to square one."

Noses

The nose is the front end of the model rocket. After this astounding and illuminating statement, let me add that its shape can vary widely. We used to call it a "nose cone," but since its shape was rarely that of a true cone, we soon started calling it just a "nose."

Some of the more common types of nose shapes are shown in Figure 4-2.

Unless your model uses a special type of recovery device such as rearward ejection (see the chapter on recovery devices), the nose must always be free to slide forward and come off. Therefore, the back end of the nose is cut down to form a shoulder, or, technically, a tenon, that will slide inside the body tube and hold the nose in place.

The base diameter of the nose should match the outside diameter of the body tube. The diameter of the shoulder, or tenon, should be slightly less than the inside diameter of the body tube (about 0.005 inch to 0.010 inch smaller) so that it has a slip fit inside the body tube. Remember, it is better to have the nose tenon a little loose because you can always build it up, or shim it, with cellophane or paper tape to get exactly the correct fit. If you have a nose tenon that is too big for the tube, it is possible to roll it down, or scrunch it, by rolling the tenon along the edge of a table, crushing the balsa down until you get the proper fit.

There are a very large number of different noses now available from

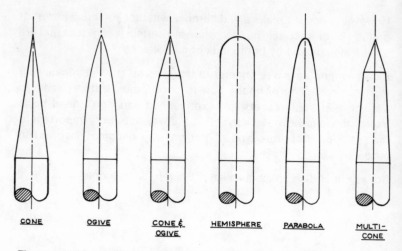

Figure 4-2: Some common nose shapes.

the model rocket manufacturers.

Noses are usually made from balsa wood or hollow plastic such as injection-molded polystyrene or blow-molded polyethylene. *Never* use a metal nose. If you must increase the weight of a nose to achieve proper stability, as we will discuss later, use a flat nose weight or, with larger noses, drill a hole in the base of the tenon and fill it up with the proper weight of glazier's lead putty.

Don't stick a metal pin or nail into the tip of a nose to simulate a probe antenna. You could have a rocket-powered dart on your hands if something goes wrong in flight. In fact, you don't even have to have a sharply pointed nose on a model rocket; slightly rounded noses work better, as we'll see later.

You can make your own noses if you have an electric drill in your workshop. Drill a hole into one end of a balsa block. Glue a 1/4-inch hardwood dowel into the block so that it protrudes about one inch. This gives you something to tighten into the chuck of the electric drill; balsa is far too soft to hold well in a simple chuck. When the glue dries so that the balsa block doesn't separate from the dowel and chase you across the workshop, chuck the dowel in the drill and proceed to carve the balsa *carefully* down to nose shape, using very coarse sandpaper, judiciously at first. This is a tough job to do. Be prepared to create some excellent egg-shaped noses at first.

The ultimate is, of course, to turn the nose on a precision lathe. You may need to do this when you get into scale model work and find yourself required to hold dimensions as small as 0.001 inch with balsa. It can be done.

Body tubes

Body tubes for model rockets are usually made from thin-walled paper tubes. Because such tubes are hard to make and even more difficult to find in hardware stores, most model rocketeers buy them ready-made from hobby stores or by direct mail from the manufacturers.

Body tubes are available in diameters ranging from 0.197 inch to 6 inches or more. Common lengths are 18 inches and 24 inches. Buy body tubes in the longest possible length and cut them to the custom length you require. You may have some scraps left over, but you will find plenty of uses for these as motor mounts, stage couplers, payload supports, etc.

To cut a tube to the required length, first measure the desired length on the tube with your steel rule and mark it. Then wrap a piece of paper or file card stock around the tube at the pencil mark and draw a line along the card edge all around the tube. Cut the tube with a No. 11 X-Acto knife blade. Take several passes around the tube, cutting only a little on each pass. Don't try to cut all the way through on the first pass around, or you will mess up the tube. After some practice you should be able to cut a body tube so that you'll never be able to tell that it was cut to custom length.

Bodies for scale models that are not cylindrical in shape can be made by the hollow log technique. The balsa body is first turned to external shape on a lathe. Once the outside shape is turned, the modeler has two choices of style. Sometimes a small hole can be drilled down the center of the turned block body so that a regular paper body tube can be inserted into the block body. The paper body tube houses the motor mount, shock cord attachment, and recovery device, like an ordinary cylindrical model. However, this approach is usually the heavy approach, resulting in a model that weighs considerably more than it would if the more refined and difficult hollow-log technique were used.

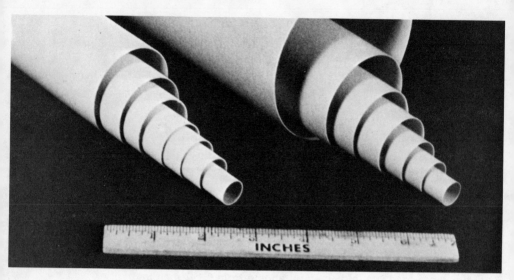

Figure 4-3: Some of the available body tube sizes. Most of them come in lengths up to 18 inches. Centuri tubes are on left, Estes on right.

For this technique, cut the turned balsa body in half lengthwise with a very thin saw blade. Hollow out both halves until the side walls are about 1/16 inch thick, just so that you can see light through the balsa. Glue a paper body tube down the middle of one of the halves. Glue the two halves back together. With a little sanding and filing you will not be able to tell where the joint is. And you will have a very lightweight, thin-walled balsa shell for a body.

Motor mounts

A motor mount in a model rocket serves two purposes: (1) to hold the model rocket motor firmly in place so that it cannot move forward under thrust or backward (and out of the model) upon activation of the ejection charge; and (2) to hold the model rocket motor straight and in alignment with the model rocket airframe so that the thrust is directed along the center line of the model.

If the motor is not held firmly in the model, the thrust can ram it forward so that it comes out of the front end of the body tube. If this happens, the motor usually reams the body tube, taking the wadding, recovery device, shock cord, and shock cord mount along with it on a short upward flight. This isn't good for your model.

On the other hand, if the motor is not mounted firmly in the model, the ejection charge can pop the motor backward out of the model. When this happens, the recovery device usually does not deploy at all. The model comes down like a streamlined anvil, usually saving the modeler the trouble of disposing of it because it buries itself. These death dives are not funny, especially since they always seem to happen during a critical demonstration when you want everything to go perfectly. So use a proper motor mount.

A motor mount usually consists of several basic parts: (1) the motor mount tube into which the motor slides with a slip fit; (2) the thrust mount ring to prevent the motor from ramming forward during thrust; (3) centering rings that will center the motor mount tube in a larger body tube; and (4) usually a thin springlike motor retaining clip.

In simple models the body tube itself has the proper diameter for the motor to slip into without the need of the motor mount tube.

A thrust mount ring is usually a small paper doughnut glued to the motor mount tube to prevent the motor from going forward. It has a hole through its center to allow the ejection charge gas to pass. Often the motor retaining clip incorporates a small tab that acts like a thrust mount.

Centering rings are often larger paper doughnuts or, for body tubes that are very large with respect to the motor mount tube, cardboard or tagboard discs. Sometimes they are slotted to clear the motor clip. Although the shock cord can be attached to them in some cases, this is not considered to be good practice because it makes the shock cord susceptible to being burned through by repeated ejection charges.

Some models and motor mounts do not use motor clips, although a clip is almost 100% insurance that the motor will not be ejected in flight; a clip also makes the prepping of the model very fast and simple. If the model does not have a motor clip, you will have to be sure that the motor is installed tightly in the motor mount tube. It should be so tight that you cannot pull it out of the model with your fingers. But you should be able to remove it by giving it a firm, sustained pull with pliers. For this reason, you should always assemble the motor mount so that about 1/4 inch of the motor

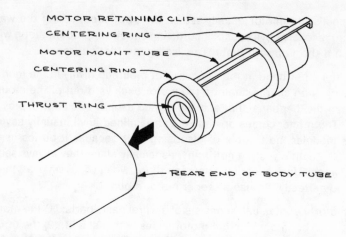

MOTOR RETAINING CLIP
CENTERING RING
MOTOR MOUNT TUBE
CENTERING RING
THRUST RING
REAR END OF BODY TUBE

Figure 4-4: Drawing of a typical motor mount for positioning and retaining a model rocket motor in a larger tube. Centering rings and motor tube can be eliminated if the body tube is the same size as the motor tube.

casing will protrude from the model so you can grasp and extract the motor with the slip-joint pliers that you naturally keep in your field box for such use. If the motor casing is flush with the aft end of the motor mount tube, you will need to *push* the motor out with a pusher rod, a 12-inch to 18-inch length of 1/4-inch hardwood dowel that you also keep in your field box—or don't you? Just stick the pusher rod down through the body tube, being careful that you don't push the entire motor mount of the model in the process of pushing out the motor. Some motors become so fond of a model that they don't want to leave.

If you do a good job of assembling the motor mount and gluing it securely in the model, it will never come out. This is an achievement worth striving for.

Fins

A model rocket must have fins, or stabilizing surfaces, on its rear end in order to fly properly. You should not experiment to determine the truth of this statement.

The fins on a model rocket are like feathers on an arrow. They keep the model going straight in the air. Model rockets do not fly in the ordinary sense of the word. Their fins are not wings that provide forces to keep them aloft, but are stabilizing devices to ensure a straight and predictable flight path.

Some of the beginner's model rocket kits have molded polystyrene plastic tail assemblies that fit right onto the rear end of the body tube. This eliminates the most difficult and time-consuming task of constructing a model rocket—getting the fins on straight and strong. If the fins are not attached correctly, the model can fly erratically or not at all. After you become an experienced model rocketeer, don't start sneering at plastic tail assemblies, however. These assemblies provide the sort of ultra-true and super-straight flights that are needed in such events as spot landing and predicted altitude. I have set an NAR national record in design efficiency with a model using a plastic tail assembly. At the time of this writing, the record still stands.

With some plastic tail assemblies it is vitally important to ensure that they do not slip off the rear end of the body tube when the thrust of the motor accelerates the model off the launch pad and into the air. When this happens—and it does to some beginners—you never want to have it happen again. The easiest way to prevent it is to wrap one or more layers of tape around the rear end of the body tube, enlarging the tube just enough for the tail assembly to have a nice, snug fit over it. Many modelers glue the plastic tail assembly to the model; this assures that it will not leave the party. But if a fin then breaks upon landing, the entire model must be thrown away because there is no way to replace the tail assembly. The best you can do is glue the broken fin back into place, and this is sometimes difficult to do.

Most model rockets, however, are built by cutting fins out of cardboard (for easy models) or balsa. They are then glued on the body tube in the proper location.

Some kit models now come with die-cut balsa fins so that you don't have to cut them out for yourself. But it is always a good idea to run around the die-cut lines with a sharp No. 11 X-Acto to ensure that the die-cut has gone all the way through the balsa sheet and that the fins will pop out of the sheet without breaking in the process.

If the fin isn't die-cut, you will have to cut it out of the sheet yourself. Model kits either have the fin planform, or outline, printed on the sheet balsa or on a paper pattern. In the latter case you must cut the fin pattern out of the paper, trace around it with a sharp pencil on the balsa sheet, and then cut the fin from the balsa with a knife.

Important: The grain of the balsa must always run parallel to the leading edge of the fin (see Figure 4-6) or *outward* from the body tube. If the grain runs parallel to the body tube, the fin will not be strong enough and will break in flight. Die-cut fins or printed-on fin patterns are already oriented with the grain parallel to the leading edge of the fin.

To make sure that you do not forget this very important point, here is a memory teaser for it: Never forget that *the grain runs out!*

Cut the fins from the sheet of balsa with your modeling knife. Try to cut square across the balsa sheet. To cut a straight line, cut along the edge of your metal rule. Making a square, straight cut may be difficult to do at first, but keep at it. Each of us develops his own personal style of holding the knife and making the cut.

Large fins can be made stronger by using thicker sheet balsa. For most small sport models 1/16-inch sheet balsa is usually strong enough. For larger models or for large fins use 3/32-inch or 1/8-inch sheet balsa. Sheet balsa in varying thicknesses from 1/32 inch to 1/2 inch is available in most hobby shops.

To further strengthen a large fin, you can glue paper to both sides of it. This also eliminates the wood grain and makes the fin easier to finish and paint. You can also cover fins in the time-honored model airplane method using tissue, silkspan, Monokote, Coverite, Solar Film, or other types of model airplane coverings. This makes a sheet balsa fin quite strong indeed.

I have built very strong fins using the polystyrene foam plastic from packing crates as a core to give the fin its shape, then covering the foam with Monokote and shrinking it to make the modeler's version of the professional aeronautical engineer's stressed skin construction. This produced a very light fin with a beautifully slick finish and excellent aerodynamic characteristics.

Once you have cut out all the fins, stack them together and sand the stack to make sure that all the fins are the same size and shape. This may not seem to be overly important with sport models, but it can become critical with high-performance competition models or with scale models. So develop this stack sanding habit from the very start in your model rocket career.

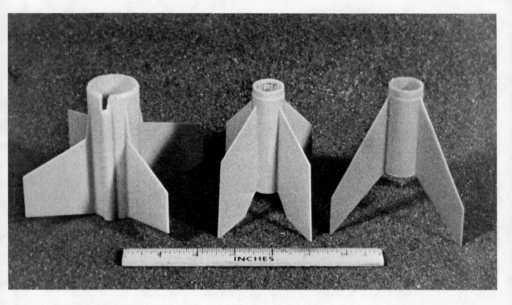

Figure 4-5: Some injection-molded plastic tail assemblies with fins integrally molded as part of the unit.

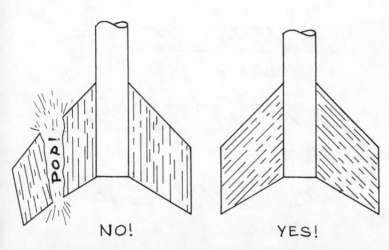

Figure 4-6: Unless you want to build weak "pop fins," always cut out fins from balsa so that the grain follows the leading edge of the fin pattern.

Your model will fly if you glue the fins on with square edges. But the performance and altitude can be nearly *doubled* if you will take the time to put a streamlined airfoil on the fins. Some common ones are shown in Figure 4-7. The simplest one merely has the edges rounded—except at the fin root, where you will glue the fin to the body tube. You can round fin edges quickly and easily with your sanding block.

In the early days of model rocketry many modelers put sharp-nosed wedge airfoils or sharp double-convex airfoils on the fins of their models, emulating the appearance of the real rockets at Wallops Station or the Cape. But the big ones are designed to fly best at supersonic and hypersonic speeds, while our model rockets rarely attain half the speed of sound. Therefore, the best airfoil for model rockets is one with a rounded leading edge and a tapered trailing edge, as shown. It takes a little more time to do this shaping with your sanding block. You may goof up a couple of fins in the process, but cut out some more and try again. The effort will pay off handsomely in improved performance.

Don't round or taper the fin's root edge that will be glued to the body

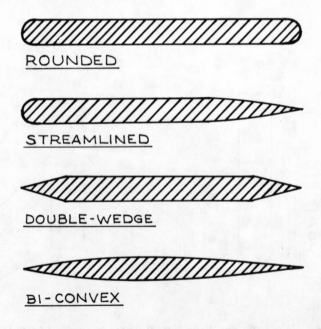

ROUNDED

STREAMLINED

DOUBLE-WEDGE

BI-CONVEX

Figure 4-7: Some basic model rocket fin airfoils shown in cross section.

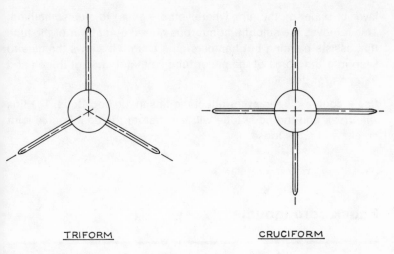

TRIFORM   CRUCIFORM

Figure 4-8: Rear view of two models showing three- and four-fin configurations.

tube! Like nearly everything else, it's happened.

How many fins should you put on a model rocket? The bare minimum is three fins in the triform arrangement shown in Figure 4-8. Most models have the four-finned, or cruciform, arrangement that is shown. Models are made with five and six fins, but rarely with more than six.

Always put the fins at or near the rear end of the body tube. The reasons for this and the reasons why fins have to be certain sizes and shapes will be discussed later.

You should *never* put the fins near the nose of the model. Nor should you put any fins anywhere except near the rear end unless the kit instructions specifically tell you to do so and tell you precisely where to put them. Things begin to get very tricky when you put the fins up front.

You can locate the fins on the body tube by using the fin placement guide as shown in Figure 4-9. Locate them carefully so they stick straight out from the body tube and are perfectly aligned with the tube. If the fins are canted or crooked or cocked, the model may take off at an odd angle, spin, fly erratically, or otherwise act up in flight. If you put the fins on straight and true, the model will fly straight and true. Don't forget that the fins are the model's stabilization system.

To get a powerful glue joint on a body tube, lightly sand off the top layer of paper on the tube where you are going to glue something. This removes the smooth, nonporous glazed outer layer of the tube that assists painting but hampers glue action. It allows the glue to seep into the pores of the paper tube so that it can "grab holt and hang on for dear life."

Use a double-glue joint for attaching fins to the body tube. The fins will break and the body tube will tear before the double-glue joint cracks and turns loose.

Shock cord mounts

The shock cord is the line that attaches the nose to the body tube. The name was first used by Orville H. Carlisle because early shock cords were quite short and had to absorb the shock of the nose flying off the front of the body tube and being brought up short. The first shock cords were rubber, but we've largely gotten away from using rubber or elastic cords now because we've discovered that elasticity isn't required. Many manufacturers still use elastic shock cords in their kits, however.

Figure 4-9: How to use the fin placement guide detailed in Figure 3-11.

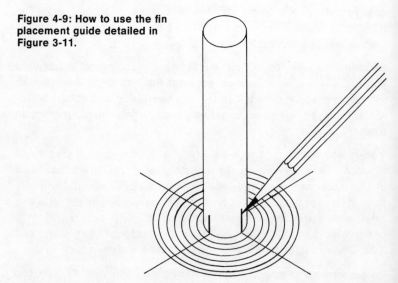

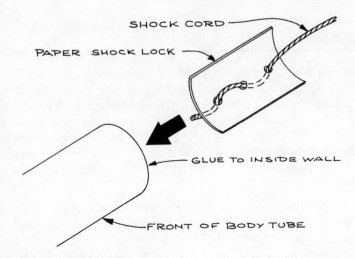

Figure 4-10: How to make and use the shock lock for installing a shock cord in a model rocket body tube.

A perfectly good shock cord is a piece of cotton kite twine no less than 18 inches long. An 18-inch cord will allow the nose to slow down sufficiently after it has popped off the model. Shorter shock cords may break. A ball of cotton kite twine purchased in a hobby store will provide you with shock cords for years of modeling! Do not use nylon, dacron, or other plastic twines; the heat from the ejection charge usually melts them. This allows the nose to separate from the body tube so that you have to recover two pieces when it is usually difficult enough to recover the whole thing tied together.

Attaching the shock cord to the body tube has always been one of the difficult problems of model rocket construction. In the early days we slit the body tube, threaded the shock cord through the slit or slits, and glued everything down flush with the body tube. But it took a lot of extra work to hide a shock cord attached like this, especially on scale models. Many other methods were tried. In 1969 I worked out the paper, or adhesive, shock lock shown in Figure 4-10. By threading the shock cord through this shock lock and gluing or attaching it inside the body tube, you don't cut or mar the tube or interfere with the ejection of the recovery package or wadding. And you have a shock cord mounting that will usually last for the life of the model. A cotton shock cord gives up from sheer fatigue after fifty to sixty flights, and you have to glue in a new one. If you can keep a model rocket flying for that many flights, you're doing very well!

Recovery devices

If you've spent time building a good model rocket, you will want to get it back after its first flight in condition to fly many more times. In addition, a 2-ounce (57-gram) model rocket falling freely out of the sky in a streamlined configuration isn't safe at all. Those are two good reasons why you should never fly a model rocket without a recovery device. Besides, it is very simple to make a good recovery device, and model rocket motors are designed to eject such a device.

All motors used in single-staged model rockets or in the upper stages of multistaged model rockets are made so that they will produce a retro-thrust puff of ejection gas at a predetermined time after ignition. This puff of gas is used to activate or deploy the recovery device. The chapter on model rocket motors explains in detail how this is done. Briefly, the motor puffs a bit of gas that pressurizes the inside of the body tube, pushing the wadding and recovery package forward, dislodging the nose, and permitting the recovery package to exit from the model and deploy in the air.

Many types of recovery devices have been developed and used by model rocketeers over the years. Some of the more successful and common ones are discussed in the chapter about recovery devices. For now, we'll take a quick overview of the most widely used ones.

The most common recovery device is a crepe paper or plastic streamer that is tied to the shock cord near the nose end of the cord. When ejected from the model, it streams in the wind and flutters, slowing the model's descent. Streamers are used only on small models weighing less than 3 ounces (85 grams). They are most widely used on small high-performance models that ascend to high altitudes or models that are flown on small fields in high wind conditions. A model with streamer recovery will return to the ground more rapidly than one with a parachute, and it will not drift so far in the wind.

A streamer is a long, narrow, rectangular strip of crepe paper (preferred for contest work) or thin plastic film. It is usually 1 to 2 inches wide and 12 to 24 inches long. A fairly standard streamer is 1 inch by 18 inches. The best length-to-width ratio is 10 to 1. A streamer should be made of brightly colored material, preferably bright orange or red, so that it can be more easily seen against the sky, on the ground, or in a tree.

The most obvious recovery device is a parachute. Most model rocket parachutes are made from polyethylene film less than 1/1000 inch thick. Commercially made model rocket parachutes are printed or decorated in bright and contrasting colors and patterns so that they may be more easily seen. A typical model rocket parachute is shown in Figure 4-12. Most model rocket parachutes have six or eight sides with six or eight shroud lines, respectively. The shroud lines are lengths of carpet thread or nylon thread. The shroud line length should be at least equal to the diameter of the parachute, and longer if possible. The shroud lines are attached by means of tape discs or strips.

Be sure to attach the shroud lines firmly to the edge of the parachute. When a parachute is ejected from a model rocket and opens in flight, it sometimes snaps full of air with a loud pop that can be heard on the ground a couple of hundred feet below. This action puts a great deal of strain on a parachute. If one or more of the shroud lines pulls off, the parachute will lower the model more quickly.

The larger the parachute, the longer and farther the model will drift in the wind. If you put a big parachute into a little model rocket powered by a high-impulse motor, you will probably never see that model again! It will drift for miles before it lands.

A parachute 8 inches to 10 inches in diameter is usually adequate for models weighing up to 2 ounces (57 grams). Models with 18-inch parachutes have set world duration records. I have lost count of the number of model rockets that I have lost with 18-inch parachutes. At

Figure 4-11: Two model rockets, one using a simple streamer recovery and the other utilizing a parachute for slower descent.

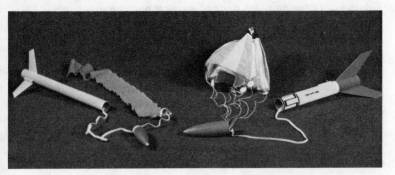

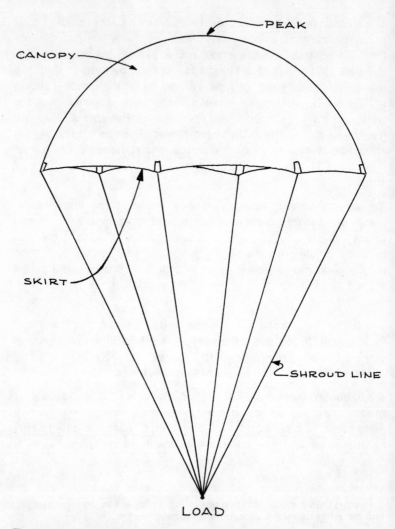

Figure 4-12: Drawing of the parts of a typical parachute.

the First International FAI Model Rocket Competition in Dubnica, Czechoslovakia, in 1966, my parachute duration competition model with an 18-inch parachute was found seventeen miles away by a Czech forest ranger who was rather astounded to discover a small rocket with American markings in the middle of the Little Tatra Mountains of Middle Europe!

Painting

Take the time and trouble to paint your model rocket, even though it may come in kit form with all parts precolored. It will look better painted—if you do a good job of painting. Usually it will also perform better if it has a smooth, shiny finish of carefully applied paint.

Model airplane dope is not usually recommended for painting model rockets, although it can be used successfully. The main problem with dope is that it is designed to shrink as it dries; this feature allows it to tighten up the tissue coverings on model airplanes. But on model rockets it often causes thin balsa fins to warp or it pulls away from the fin-body joint into a fillet that is larger than you need or want. In addition, model airplane dope also melts, or crazes, the surfaces on polystyrene plastic parts.

Hobby stores carry two types of paint that can be used on model rockets. These are *enamels* and *acrylics*. Testor's and Pactra both make enamels in a variety of colors. However, enamels straight out of the bottle or can are usually too thick, leave a lumpy finish, and yellow or change color with age. Most model rockets painted with enamel by beginners look pretty bad. I often think that some modelers smear the enamel on with a cotton swab or a toothbrush. Acrylic polymer paints made by Floquil and others can be thinned with water, go on smoothly without showing brush strokes, do not yellow with age, dry to the touch in 30 to 60 minutes, produce a hard, water-resistant finish, and don't smell.

The finest model rocket paint job is one applied by spraying. Enamels, alkyd enamels such as Krylon, and other types of paints are available in a wide variety of colors in aerosol spray cans ranging in price from under a dollar for a small can to several dollars for a large can. To use these aerosol spray paints, put your model on a paper wand and give it several thin coats of paint for the best results. Do not try to get a complete, all-covering paint job in one application; the paint will run, producing an unsightly glob. "Dust" the paint onto the model. You can put on several light coats and have them dry in less time than it takes for one thick, globby coat to run and dry.

Small hobby-type airbrushes are now available at reasonable cost in hobby shops. These are ideal for painting model rockets. The Stines have several of them in the shop and have actually worn out two. Water-based acrylic paints are best to use in airbrushes because you

Figure 4-13: An airbrush provides a beautiful, even coat of paint on a model rocket and is well worth the expense. Be sure to use lots of newspaper to prevent overspray from getting everywhere.

can thin the paint with water and clean out the airbrush with water when you change paint colors or get ready to put things away for the night. Pressure for airbrushes usually comes from cans of pressurized Freon gas sold in the hobby stores. These are all right for quickie paint jobs, but you will soon grow tired of buying Du Pont's Freon cans. The next best bet is to buy a small air compressor at Sears, Ward's, or the hobby shop. A compressor will set you back a fair piece of change, but it will soon pay for itself if you do a lot of spray painting. It is also handy for inflating bicycle tires, blowing sawdust around, etc. Some airbrushes will work on the output of a tank vacuum cleaner; attach the hose to the end of the tank vacuum cleaner that blows instead of sucks. For use in very damp or humid climates or for portable use, you can buy small cylinders of pressurized carbon dioxide (CO_2), which, when used with a pressure regulator to drop the pressure, will produce a steady stream of dry gas for airbrushes.

Paint your model rocket in bright colors that can be easily seen in the air and on the ground, preferably fluorescent red or orange, as suggested earlier. A model rocket is most highly visible if it is painted in one solid color with perhaps one black fin. Painting it many colors will tend to camouflage it and make it hard to see.

Fluorescent colors are available in the hobby stores in spray cans or in hardware stores in big Krylon cans. You must spray on fluorescent colors; painting them on with a brush leaves them very streaky in appearance. Fluorescent colors must also be applied over a base coat of flat white paint.

The best nonfluorescent colors for maximum visibility depend upon the general sky color and condition in your locale. Against the usually clear blue skies of the American West and Southwest, the best colors are white, orange, or red. Against the often cloudy, gray, hazy skies of the Midwest, South, and East Coast, dark colors such as black or maroon work well. Some silver paints show up well against cloudy or gray skies, but they often wash out by taking on the sky color itself; then they tend to disappear against the sky. Greens, browns, and sky-blues blend too well with the surrounding environment. One not-very-bright model rocketeer I once knew painted his model in accurate military camouflage colors that really worked; he could not find the model after it landed on the ground!

Decals and trim

Most competition models don't carry much decoration other than a decal or marking showing the contestant's sporting license number as required by national rules. Competition modelers have learned that putting decals and strongly contrasting patterns on a model tends to make it more difficult for the human eye to follow. Note that military camouflage and the natural color of animals uses a number of contrasting colors in rather random-yet-regular patterns.

However, the application of decals and trim to sporting models results in a striking change in their appearance. The smooth, streamlined, uncomplicated shapes and lines of the model suddenly seem to come alive; the model begins to look more like its big brothers at the Cape. To obtain a good-looking model that resembles a real rocket, it is important to know what kind of decals and trim to use. This is because the color patterns and markings on the real ones have definite functions.

The most important decal or marking is the roll pattern of regular stripes, checks, or other marks around the body. At the Cape, roll patterns are painted on the vehicles for photographic purposes so that engineers can determine from camera films how the rocket rolled or tilted in flight. If you will roll one of your models in your hand, you will see how the roll pattern decal changes appearance and how it is possible to determine roll rate from that appearance.

Usually one fin is painted black or some contrasting color. This is to provide a roll reference point. Fins sometimes have different paint patterns on them so that engineers can tell from photographs just which fin they are looking at.

Most NASA space vehicles and sounding rockets carry a painting of the American flag and the words "United States" or "USA" on the body of the first stage. The star-and-bar military insignia is never applied to NASA rockets. That insignia, however, and the words "U.S. Air Force" or "USAF" usually are placed somewhere on the body of an Air Force rocket. Often USAF rockets also have some sort of emblem, squadron, command, or other type marking on them.

Numbers are usually applied to the fins but may also appear on the body. Numbers and letters are usually applied so that they can be read when the rocket is in the launching or flight attitude.

Follow the instructions on the back of the decal sheet for applying the decals. Always blot a decal, especially the large roll patterns, to get the air and water bubbles out from under them. A decal should be allowed to dry for an hour or more. Otherwise, it may rub off the model.

Other markings may be applied to a model using an India ink pen such as the Mars-700 or Rapidograph.

Kits and supplies for making your own decals are available in hobby stores. Check in the craft department. Or check in an art supply store.

After all decals and markings have been applied and are dry, it is best to apply one light coat of dull or gloss transparent spray over the entire model, depending on whether you want it to have a flat or shiny finish. Dulling spray will kill the shine of the decals and make the model look very big and very realistic.

Figure 4-14: The proper placement of a few simple decals on a model rocket improves its appearance tremendously!

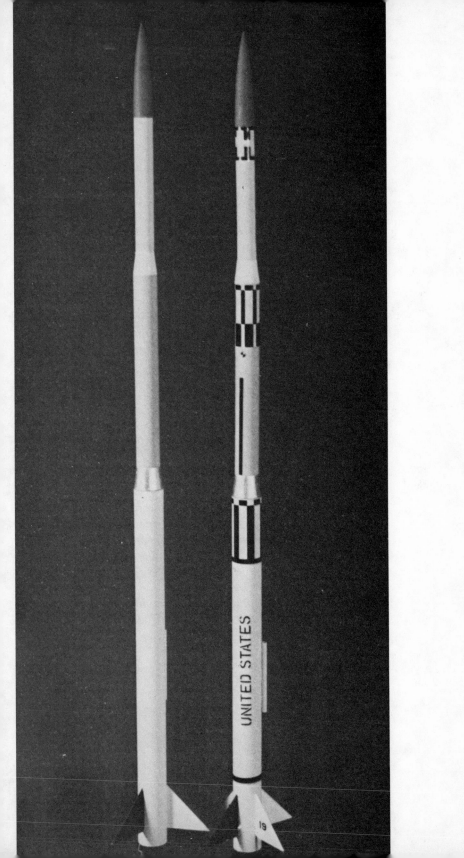

Storage and carrying

Although your model rocket may be strongly constructed and capable of flying at half the speed of sound through the air, it is surprisingly easy to break when it is on the ground. Little Brother may get to it. Fido may decide it is a superior and very tasty bone. Big Clod may step on it. Fumblefingers may drop it three feet to a concrete floor where it will smash into ten thousand little bitty pieces, most of which you will never find again. Or it simply may get trashed by an overzealous Mom who doesn't understand that you are trying to get one hundred flights on that ratty old bird.

Put your models away between flight sessions. Make a display rack for them. Or hang them from the ceiling by thread. Or put them in a big drawer—but be careful not to shut the drawer on the fins! A big cardboard box is great for storing model rockets.

To keep a model rocket from being damaged in storage and during transit to the flying field, put it in a plastic bag. The air trapped inside the plastic bag with the model acts like a cushion to protect the model. I have carried model rockets back and forth across the Atlantic Ocean in plastic bags inside a cardboard box without the slightest damage or paint smudge. This is why I always open the plastic bag of a kit model very carefully; I save the bag and put the completed model in it.

Now that we have covered some of the basics of model rocketry, we can talk about some of the more advanced parts of the hobby. Before we go on, however, you'd be smart to reread these first chapters again, just to be sure that you understand what we've been talking about and will remember most of it. Even after you've read about some of the more complex aspects of model rocketry, come back and review these fundamentals from time to time. In that way you'll have a sound understanding of the fundamentals and won't be held back by making basic mistakes.

Model Rocket Motors

The device that makes model rocketry a hobby and a sport rather than a disaster is the model rocket motor. Some people call it a model rocket "engine," but the distinction between "motor" and "engine" is subtle. A motor is defined as "something that imparts motion," while an engine is defined as "a machine that converts energy into mechanical motion." Thus, a steam engine or an internal combustion engine is definitely an engine. A model rocket motor is truly a motor. Technically, it is a small reaction device for converting the energy of high-temperature gas into motive power without the use of gears, cams, linkages, pistons, turbines, etc.

Model rocket motors are available in two basic types—the solid propellant and the cold propellant. Cold propellant model rocket motors are used mainly in locales where solid propellant model rocket motors are still restricted by archaic laws adopted long before the space age dawned. The solid propellant model rocket motor is the type most widely used by model rocketeers in the United States and abroad, so we'll spend most of our time discussing it.

Solid propellant model rocket motors

A solid propellant model rocket motor is an inexpensive, highly reliable package of power that comes all ready for use. It will provide

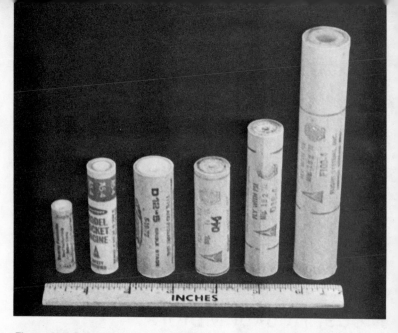

Figure 5-1: Some of the model rocket motors manufactured in the United States at the time of this writing. Units shown are representative of the different sizes available.

both the propulsive force to thrust a model rocket hundreds of feet into the air and the means for ejecting the recovery device. Model rocket motors are the world's most reliable rocket motors; by 1974 over 50 million had been manufactured and used without producing any personal injuries more serious than a burned finger and without creating any severe fire hazard.

A model rocket motor appears to be a very simple device, but it is actually very complicated. Making one costs a large amount of money and requires a great deal of expensive and very special equipment, plus a lot of knowledge. Strict safety precautions must be taken to manufacture a model rocket motor. This is why model rocketeers leave the making of rocket motors to professionals, the model rocket manufacturers.

There is no safe way to make a rocket motor of any type. This is a statement of fact, not a matter of opinion. There is no way that you can make a rocket motor that will be as inexpensive and reliable as a commercial motor. A model rocket motor is a factory-made device that is subjected to rigid quality standards and rigorous statistical batch testing procedures. It is very reliable and will do exactly what it was designed to do. You, as a model rocketeer, do not have to handle dangerous chemicals, worry about whether or not the motor will have proper thrust, or take extensive and expensive safety precautions.

Never forget, however, that a model rocket motor is *not a toy*! You should understand this right from the start. If you do not use a model rocket motor correctly and in accordance with instructions, you could get hurt by it. This is also true of most technical devices. A model rocket motor is packaged propulsive power for models, and should be used for no other purpose. It should be your introduction to the fact that technology can work for you if you handle it right—and that it can bite you if you don't. Cro-Magnon man learned about technology the hard way when he got burned by his cave fire. Fortunately, you don't have to learn everything in that school of hard knocks called experience, but you must be willing to learn.

As of this writing there are sixty-one different types and makes of model rocket motors available in the United States. From this broad selection—the greatest available in any country in the world—you should be able to choose a model rocket motor that meets your requirements.

A solid propellant model rocket motor is shown in Figure 5-2 as though it were cut down the middle to expose the innards. (Don't do it! It's safer to look at the drawing!)

The motor *casing* is made from tightly wound kraft paper with carefully controlled dimensions. Paper is used because it is very strong, is fire-resistant, and does not conduct heat easily. The dimensions of model rocket motor casings vary according to the power, manufacturer, etc. However, there are some basic model rocket motor sizes, as shown in Table 4.

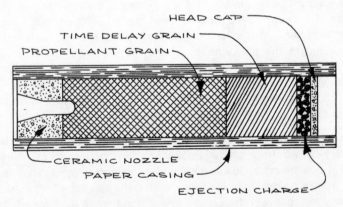

Figure 5-2: Cross section drawing of a typical solid propellant model rocket motor.

Table 4
Model Rocket Motor Size Chart
Correct and complete as of January 1975

Family or Series	Manufacturers	Diameter × Length	
		(millimeters)	(inches)
"Mini"	Centuri and Estes	13 × 45	0.50 × 1.75
Standard	Centuri and Estes	18 × 70	0.69 × 2.75
Estes D	Estes	24 × 70	0.945 × 2.75
FSI A through D	Flight Systems	21 × 70	0.83 × 2.75
FSI D18 and E	Flight Systems	21 × 95	0.83 × 3.75
FSI F	Flight Systems	27 × 150	1.06 × 5.91

The *nozzle* is made from ceramic material formed into a carefully designed size and shape. You should never try to alter the nozzle because the slightest change in dimension of the nozzle throat—as little as a thousandth of an inch—could drastically alter the operation of the motor.

The *solid propellant* is a rock-hard piece of combustible chemical material with controlled burning characteristics. You needn't know what the chemical composition of the propellant is; you are far more interested in the performance of the motor. Once ignited on its nozzle end, the propellant burns forward from the nozzle, producing over two thousand times its solid volume in hot gas. This gas shoots out of the nozzle to produce thrust in accordance with Newton's Third Law of Motion. More about this later.

Ahead of the solid propellant is the *time delay charge*. This is a piece of very-slow-burning propellant that produces practically zero thrust and allows the model to coast upward on its momentum to apogee. If there were no delay charge in the motor, the ejection charge would deploy the recovery device at a low altitude and a high speed. The model and its recovery device are not strong enough to withstand this sort of flight behavior more than once.

The time delay charge lasts several seconds, depending upon the type of motor you install in the model before the flight. The end of burning of the time delay automatically activates the *ejection charge*. The ejection charge produces a quick puff of gas that pressurizes the inside of the model rocket, pushes the recovery wadding and device forward, dislodges the nose, and expels the recovery device from the model. The ejection charge may be held in place with a *head cap* that is either a paper cap or a thin ceramic plug that is shattered when the ejection charge is activated.

And that is all there is to it.

And it works.

You should never attempt to reload a used solid propellant model rocket motor casing. It has been designed for only a single firing. If you want to cut apart a *used*, expended motor casing (never a loaded one), you will see that the inner layer of the casing has been charred. This has probably weakened the casing so that it would not be safe to use again. In addition, as pointed out earlier, the handling of rocket propellants and the making of rocket motors are jobs for trained experts—and you aren't one.

You will find a lot of information printed on a model rocket motor casing and even more in the package and instructions that come with the motor. The most important thing for you to check for is the statement "NAR Certified." This means that the Standards and Testing Committee of the National Association of Rocketry, the nonprofit spokesman for model rocketry in the United States, has tested samples of that type and make of motor and has determined that the motor type meets or exceeds a strict set of performance standards jointly developed by the model rocket manufacturers who are members of the Hobby Industry Association (HIA) and by every-day model rocketeers like you and me who are members of the NAR. If a model rocket motor doesn't have "NAR Certified" printed on it, its instruction sheet, or the box in which it came, you should be wary of it. In some states it is against the law to sell or use a model rocket motor that does not have the Safety Certification of the NAR.

All NAR-certified model rocket motors carry on their casings the universal United States model rocket motor code that tells what kind of motor it is and how it will perform. This NAR motor code is simple. It consists of a letter, a number, a dash, and a final number.

A typical example might be: B4-6.

The first *letter* of the code indicates the power range of the motor.

Figure 5-3: A static test of a model rocket motor being conducted by the NAR for safety and performance certification.

How do you figure the power of a model rocket motor? In terms of horsepower? Starpower, maybe? No, in terms of *total impulse*, which is a factor derived by multiplying the average thrust by the thrust duration. Or, more accurately, total impulse is the area under the thrust-time curve.

Sound confusing? Well, it isn't if you take it one step at a time.

The *thrust* of a motor is the amount of force, or push, produced when it is operating. The jet of supersonic gas rushing out of the motor nozzle produces a force according to Sir Isaac Newton's Third Law of Motion. Stated simply in words, this is: For every acting force, there is an equal and oppositely directed reacting force. Written as an equation, the universal shorthand of science, this is: $MA = ma$.

The thrust of a model rocket motor is rarely constant. It changes with time. Therefore, we must further define which thrust level we are speaking of. There are several. *Maximum thrust* is the highest amount of force produced by the motor during operation, regardless of when that occurs during the period of operation. *Average thrust* will be defined in a moment.

Because model rocketry is an international sport, motor performance specifications are in the units of the international metric system—more specifically, in terms of the MKS system. This is the meter-kilogram-second system. Motor thrust is therefore given in newtons, named after the aforementioned Sir Isaac Newton. A newton is defined as the force required to accelerate 1 kilogram (2.2 pounds) of mass at a rate of 1 meter per second per second (3.28 feet per second per second). It is easy to convert newtons of force into pounds of force:

1 pound of force = 4.45 newtons

The model rocket motor begins to produce thrust at the instant *ignition* occurs. *Burnout* is the instant that the motor ceases to produce measurable thrust. The length of time thrust is produced is called *duration*, and it is measured in seconds of time. A duration of 2 seconds is very long for a model rocket motor. The time interval from ignition to maximum thrust is known as *T-max*. This is an important parameter to know if your model is heavy; it will help you choose the right launch rod length to use, as we'll see later.

To accurately determine thrust, maximum thrust, duration, T-max, and other performance characteristics, a model rocket motor must be given a static test. It is fastened into a static test stand with recording devices attached, then is ignited and operated.

The data record of a static test produces a thrust-time curve such as is shown in Figure 5-4. This is a typical thrust-time curve for a model rocket motor. Notice how the thrust rises rapidly after ignition to a high maximum thrust. This accelerates the model rapidly to high speed, ensuring that it has sufficient airspeed before leaving the launch rod for the fins to be able to stabilize it. Then the thrust level settles back to a lower value, a sustaining thrust that accelerates the model to higher speeds during the climb.

When we static test a model rocket motor, it is no longer usable for flight. How do we know that other motors of its type are going to produce the same performance? Answer: We don't know for sure. We can only infer that they will, but that inference is fairly confident, based on statistical sampling and proven testing techniques. You can delve into this area of math in detail if you are a math shark—and a lot of model rocketeers are.

Figure 5-4: Typical thrust-time curve of an end-burning solid propellant model rocket motor showing various events.

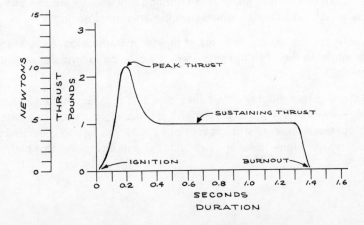

Each model rocket manufacturer has his own high-precision model rocket static test stand. They are a far cry from the simple stands using springs and rotating ice cream drums that were common in the early days and are still common in many high schools today. Manufacturers subject a random sample, usually 2%, of each motor production batch to a static test to check their conformance to NAR standards. If the test sample fails, the entire production lot is rejected and destroyed unless corrections can be made. Over a period of time the manufacturer gains a great deal of information from thousands of static tests. He can have a high degree of confidence that the rest of the motors in the batch will perform the same way. After all, if there have been 50 million motors made and used, this means that there have been 1 million static tests made!

In addition, the NAR Standards and Testing Committee obtains and tests random samples of model rocket motors from all sources. Committee members check to see that production motors in stores are the same as the original test motors and that the manufacturers are maintaining their quality standards.

As stated earlier, thrust and time, or duration, are equally important in determining motor performance. It does no good whatsoever to have a thrust of 100 pounds (445 newtons) if it lasts for only 0.01 second! Likewise, a thrust of only 0.1 pound (0.445 newtons) lasting as long as 10 seconds probably won't even lift the model into the air! The performance factor that really determines the flight characteristics of a model rocket is the *total impulse* of the motor.

We can determine total impulse from a thrust-time curve by multiplying the thrust by the duration. However, thrust isn't constant, so we usually determine total impulse by a more precise method: measuring the area under the thrust-time curve—which is the same thing, isn't it? We can do this by laying the thrust-time curve chart over a sheet of quad-ruled paper and counting squares. Or we can use a plane polar planimeter to measure the area more accurately.

The result comes out in terms of thrust-times-duration, or newton-seconds. In the old English system it would be in terms of pound-seconds.

Total impulse roughly determines how fast and how high a model rocket will fly. Flight calculations discussed in detail in a later chapter will show why this is so. For now it is enough to know that a motor with higher total impulse will take a model rocket to a higher altitude.

Table 5
Model Rocket Motor Total Impulse Ranges
United States Model Rocket Motors,
NAR Standards

Type	Total Impulse (newton-seconds)	Total Impulse (pound-seconds)
1/4A	0.00 – 0.625	0.00– 0.14
1/2A	0.626– 1.25	0.15– 0.28
A	1.26 – 2.50	0.29– 0.56
B	2.51 – 5.00	0.57– 1.12
C	5.01 –10.00	1.13– 2.24
D	10.01 –20.00	2.25– 4.48
E	20.01 –40.00	4.49– 8.96
F	40.01 –80.00	8.97–17.92

This is why the NAR rates model rocket motors in classes based on their total impulse. The NAR total impulse classifications are shown in Table 5. All American manufacturers use these classifications.

Thus, the first letter in the NAR motor code tells you the total impulse range of the motor. Most motors perform at the top of their range, although some of the larger motors may fall in the middle of a range. However, you need not look at the exact range figures of a motor to have some idea of its performance, for each total impulse class is roughly double that of the previous class. This means that a Type B motor will have about twice the total impulse of a Type A, and that a Type C motor will have twice the total impulse of a Type B and four times that of a Type A.

The *first number* in the NAR motor code (Type B4-6 in our example) tells you the *average thrust* of the motor. This is a derived, or

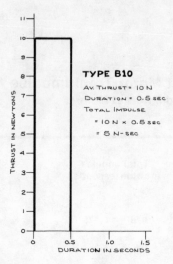

Figure 5-5: Thrust-time curve of a hypothetical Type B10 motor.

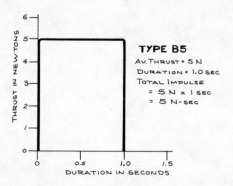

Figure 5-6: Thrust-time curve of a hypothetical Type B5 motor.

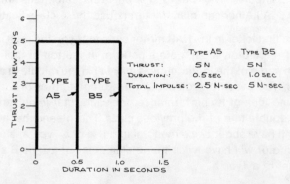

Figure 5-7: Thrust-time curves of hypothetical Type A5 and B5 motors showing equal thrusts but different total impulses.

calculated, number. It is determined by dividing the total impulse by the thrust duration. It indicates what the thrust would be if it were constant from ignition to burnout. Average thrust is a useful piece of information for altitude prediction; it also tells you how fast your model will accelerate into the air and roughly how fast it will be going when it leaves the launch pad.

We can clarify this by looking at Figures 5-5, 5-6, and 5-7.

Figure 5-5 shows the ideal thrust-time curve of a hypothetical Type B10 motor with an average thrust of 10 newtons for a duration of 0.5 second. This is a high-acceleration motor useful for flying heavy models or for flying regular models in high winds to prevent weathercocking; more about this later.

The Type B5 motor whose thrust-time curve is shown in Figure 5-6 has an average thrust of 5 newtons and a thrust duration of 1.0 second. The Type B5 motor would be a better choice for small, lightweight high-altitude models.

Both the Type B10 and the Type B5 have the same total impulse—5 newton-seconds (abbreviated N-sec). But their thrust characteristics are quite different, and they would make models perform quite differently.

Figure 5-7 shows the thrust-time curves of two motors with identical average thrusts but different total impulses. The Type A5 has 5 newtons of thrust and a duration of 0.5 second. The Type B5 has 5 newtons of thrust and a duration of 1.0 second. The Type B5 will obviously take a model to a higher altitude because the thrust will be applied to the model for a longer period of time.

The *number following the dash* in the motor code tells you the number of seconds after burnout before the motor activates the ejection charge and deploys the recovery device.

Thus, our example Type B4-6 motor has a total impulse in the range of 2.51 to 5.00 N-sec, an average thrust of 4 newtons (a little less than 1 pound-force), and a time delay of 6 seconds.

This is nearly everything you need to know to select the proper motor for your model rocket. You really can't get by with less basic information. This system has been in use since 1960. There have been many attempts to simplify it, but nobody has yet been able to devise a coding system that is so easy to remember but still tells all the basic information you need to know about a motor. If anybody says the system is too complicated for them to learn, it's my guess that they probably don't have enough gray matter to be involved in model rocketry anyway!

You must match the motor to the model before the flight. You must select the proper time delay. In the beginning follow the recommendations of the kit manufacturer regarding time delay. Later you will have enough experience to be able to decide for yourself which delay to use.

If the time delay is too short, the model will deploy its recovery device while it is still climbing. This may happen at high airspeed, and the force may tear the recovery device to shreds or rip it off the model.

If the time delay is too long, the model will climb to apogee, arc over, and begin to fall back toward the ground. It will gain speed as it falls. Again, the recovery device may be deployed at high speed and tear itself to pieces; the model may also be destroyed in the process. Cliff-hanger flights are no fun, and they could be hazardous.

Rule of thumb: When in doubt, use a shorter time delay. If the recovery device activates going up, use a longer delay on the next flight.

Some model rocket motors are dash-zero types. That is, the motor code ends in a zero: Type B6-0, for example. These motors have no time delay or ejection. They are intended for use in the lower stages of multistaged models. A later chapter is devoted to multistaging.

Another rule of thumb: Don't overpower your model. First flights should be made with the lowest total impulse recommended by the kit manufacturer. If you overpower your model, you are likely to lose it on the first flight. The kit manufacturer will love you because you will have to buy another kit. The motor manufacturer will love you, too, because the high-powered motors have more propellant in them and so cost more money.

You can often estimate how high a model will go; it is even possible (as we will see) to predict with great accuracy the precise altitude that will be achieved. For starters, you can expect the following performance out of a basic tyro model weighing 1 ounce (about 30 grams), using a 3/4-inch (20 millimeters) body tube, and equipped with these motors:

Type A8-3: 500 feet (150 meters)
Type B6-4: 1,000 feet (300 meters)
Type C6-6: 1,600 feet (500 meters)

The time delays shown in the motor codes are about right for a 1-ounce model.

As you can now see, a model rocket motor, although appearing to be a simple device, is really fairly complex. It is the result of decades of development. Each motor type took months of work to bring it up to the level of performance and reliability that would permit it to be sold to the public. Many thousands of dollars were required to buy or make the special equipment necessary to manufacture the motors. Often this equipment is hazardous to operate. A great deal of credit must be given to the model rocket motor manufacturers, who undertake grave risks to make that simple paper-cased power package. Treat it properly and it will do some rather amazing things for your models.

Cold propellant model rocket motors

In 1969 a new type of rocket came on the market propelled by a radically different type of model rocket motor. Alan Forsythe and Carl Terse of Vashon Industries produced a model rocket that was not propelled by a combustible propellant and was, therefore, not subject to some of the old restrictive model rocket and fireworks laws that were still in existence at that time.

Instead of a solid propellant that undergoes an exothermic, oxidizer-fuel, combustion-type chemical reaction to provide the large volumes of high-pressure gas necessary for rocket reaction thrust, the cold propellant model rocket motor uses a liquefied refrigerant gas technically known as dichlorodifluoromethane. It is popularly called Freon-12, the registered trade name given it by E.I. du Pont de Nemours, who developed it as a refrigerant. It is also used as a pressurizing gas for aerosol spray cans.

Du Pont's Freon-12 is a colorless, almost-odorless gas with a boiling point of -20° F. at atmospheric pressure (14.7 pounds per square inch, or psi). If kept at a pressure of 85 psi, it is a liquid at 68° F., or room temperature. If the pressure is removed and the Freon-12 in a liquid state is introduced into the atmosphere, its temperature immediately drops to -20° F., and it boils at once into a gas. This is why it gets so very cold. It cannot exist as a liquid at atmospheric pressure unless it is at -20° F.

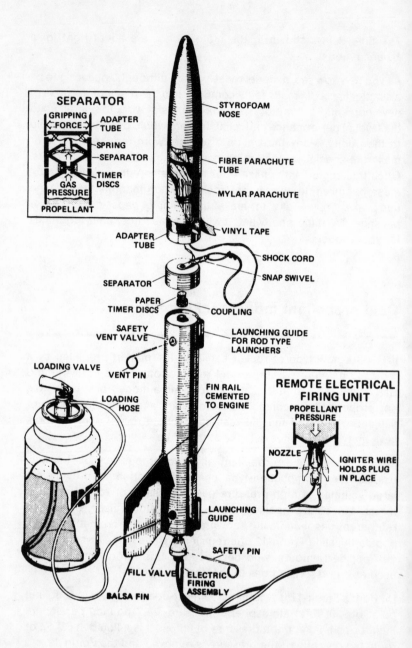

SEPARATOR

GRIPPING
FORCE
ADAPTER
TUBE
SPRING
SEPARATOR
TIMER
DISCS
GAS
PRESSURE
PROPELLANT

STYROFOAM
NOSE

FIBRE PARACHUTE
TUBE

MYLAR PARACHUTE

VINYL TAPE

ADAPTER
TUBE

SHOCK CORD

SNAP SWIVEL

SEPARATOR

PAPER
TIMER DISCS

COUPLING

SAFETY
VENT VALVE

LAUNCHING GUIDE
FOR ROD TYPE
LAUNCHERS

LOADING VALVE

VENT PIN

LOADING
HOSE

FIN RAIL
CEMENTED
TO ENGINE

**REMOTE ELECTRICAL
FIRING UNIT**

PROPELLANT
PRESSURE

NOZZLE

IGNITER WIRE
HOLDS PLUG
IN PLACE

LAUNCHING
GUIDE

SAFETY PIN

FILL VALVE

ELECTRIC
FIRING
ASSEMBLY

BALSA FIN

**Figure 5-8: Cutaway sketch of a typical cold propellant model rocket
motor.**

A typical cold propellant model rocket motor is shown in Figure 5-8. Basically, it is just an aluminum tank 1 inch in diameter and 4 to 8 inches long. There is a nozzle at the bottom end. In some types of cold propellant motors there is a separate filling valve; in others the motor is filled through the nozzle. This "fuel tank" also contains a pressure relief valve that will release the internal pressure if it should exceed 100 psi. On the upper, or forward, end of the motor is the time delay and recovery ejection mechanism. This is quite ingenious, and we'll talk about it in a moment.

In some cold propellant model rockets the motor is slipped inside the model's body tube just like a solid propellant motor with the exception that it is more permanently mounted than the solid motor; the cold propellant motor does not have to be changed after every flight. In other types of cold propellant models the motor tank doubles as the model's body tube; fins and other parts are mounted directly to the tube with contact cement or epoxy.

The motor tank is made from aluminum. It is the *only* major structural, load-bearing component in model rocketry that *must* be fabricated of metal. It is the *only* such part that is permitted to be metal. The tank cannot be made of paper or cardboard because the liquid Freon-12 seeps through the pores of the paper under pressure. Nor can it be made from plastic such as polystyrene because liquid Freon-12 embrittles and crazes plastic, making it very susceptible to shattering into sharp fragments under pressure at low temperatures.

The Federal Aviation Administration (FAA), acting upon the technical advice of the NAR Standards and Testing Committee, adopted a special waiver to the no metal rule that prohibits metal model rockets under federal law. It is permissible to fly metal cold propellant model rockets if they do not exceed the 16-ounces gross weight and 4-ounces of propellant limits established by the FAA.

A cold propellant motor operates in the following manner:

The model is positioned on the launcher with a launching assembly locked in place in the nozzle. This assembly is nothing more than a fancy sort of cork that seals the nozzle and is equipped with a quick-release that will free the cork when the rocketeer pulls the launching lanyard. From a pressurized can liquid Freon-12 is loaded into the motor either through a filling valve or through the nozzle cork. The V-1 motor from Estes is rated as NAR Type A8-CP (the letters after the dash signifying cold propellant); it holds 2 ounces of liquid Freon-12. The Estes V-2 motor is rated as NAR Type B8-CP and holds 4 ounces of Freon-12. It is important to get the motor full of Freon-12.

When you can't get any more Freon-12 to flow into the motor, the model is ready to launch. Stand back to get your head out of the way—and pull the lanyard. This releases the cork from the motor nozzle. Under a pressure of 70 psi (relative to the atmosphere) the liquid Freon-12 will immediately start to squirt out of the nozzle. However, as it does so, its pressure immediately drops to atmospheric pressure. This drops its temperature, causing it to flash over to a gas.

The gas produces the thrust. Estes' current cold propellant motors have a thrust that rises almost instantly to 2 pounds when the model is launched at 70° F. If you try to fly the model on a cold day with an outside temperature of 40° F., its thrust will be only 1 pound. If you tried to launch at -20° F., the Freon-12 would not boil at all, and the motor would have no thrust whatsoever! Obviously, then, thrust and total impulse are dependent upon the temperature of the outside air.

Thrust lasts for about 0.75 second with the Type A8-CP and about 1.5 seconds for the Type B8-CP. Duration is independent of temperature under normal flying conditions.

A typical thrust-time curve of an Estes Type B8-CP motor is shown in Figure 5-9.

The cold propellant motor, however, is not a simple blow-down motor like a Park pressurized water rocket that you just pump up and

Figure 5-9: Thrust-time curve of an Estes NAR Type B8-CP cold propellant model rocket motor.

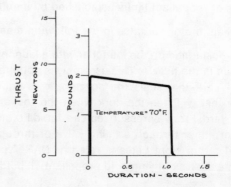

release. At launch there is a pocket of gaseous Freon-12 on top of the liquid Freon-12 in the motor tank. This is at a pressure of 70 psi when the ambient temperature is 68° F. The instant that the liquid Freon-12 starts to blow out of the nozzle, the pressure on top of the Freon-12 load starts to drop. This pressure drop in turn causes more Freon-12 on top of the motor to boil off in order to maintain the head pressure. This situation exists for the entire duration of thrust. The head pressure on the Freon-12 in the tank remains steady at 70 psi until the last Freon-12 goes out of the nozzle. The pressure then drops at once to zero, and so does the thrust. This is why a cold propellant motor can maintain nearly constant thrust except for a slight loss due to general cooling of the entire motor during operation.

Because the gas coming out of the nozzle is very cold, the propulsive efficiency, called specific impulse, is not very high for a cold propellant motor. In fact, it is only about one-tenth that of a solid propellant motor. Therefore, it takes about ten times as much cold propellant to produce the same total impulse as a comparable solid propellant motor. Because of this difference, a cold propellant model is heavier at lift-off, and most of this additional weight is Freon-12 propellant. Typically, a cold propellant model with 4 ounces of Freon-12 in it weighs about 6 to 7 ounces at lift-off. Usually such a model generates 32 ounces (2 pounds) of thrust, making the lift-off very slow because the thrust-to-weight ratio is only about 4. However, the model rapidly uses up the propellant load, so just before burnout the thrust is about 12 times the weight. Therefore, cold propellant model rockets perform more like the big ones at the Cape. They take off more slowly and are really perking along at burnout.

The exhaust from a cold propellant motor is nontoxic but very cold. Don't stick your fingers in it; they might get frostbitten. By all means, do not look into the nozzle or squirt Freon-12 around; you could get it in your eyes and literally freeze your eyeballs!

The time delay function and ejection on cold propellant motors are excellent examples of clever engineering. The time delay is actually a slow leak. On top of the motor tank is another smaller tank connected by a small tube. In this tube, or coupling, are a number of tiny paper discs or an extremely small hole. When the motor is being filled with Freon-12 on the launch pad, gaseous Freon-12 seeps through this controlled leak into the smaller upper tank. It may take 3 to 5 seconds to leak out again once the motor has reached burnout

Figure 5-10: Total absence of heat and flame permits model rocketeers to be close at launching of a cold propellant model rocket.

and the internal motor tank pressure has dropped to zero (atmospheric pressure).

When there is pressure inside the smaller time delay tank, the pressure holds one or more springs against the model's recovery section, pressing it firmly to the model. When the pressure in the delay tank drops to zero, the springs are relaxed, and the recovery section is free to come off. It is either stripped from the model by the air flowing past on the outside, or it is physically separated by another spring. This deploys the recovery device to lower the model gently to the ground.

The Estes V-1 and V-2 motors have variable time delays that you can set before flight by inserting the proper number of tiny paper discs and thus controlling the leak rate. The Estes Cold Power Convertibles have preset time delays that cannot be changed.

Cold propellant motors must not be loaded with anything but the Freon-12 available from Estes especially for cold propellant motors. Like solid propellant motors, these motors must not be tampered with or modified. Even though they do not operate on the principle of combustion at elevated temperatures, they are high-pressure devices that have liberal safety factors incorporated into their design. But no manufacturer can protect an idiot from himself. It is possible to get hurt with a cold propellant motor just as it is with a solid propellant motor—or with a baseball bat, for that matter.

A rocket motor of any type must always be treated with caution and respect.

For a number of reasons cold propellant model rocket motors have not enjoyed the wide popularity of solid propellant motors. There is a great deal more involved in the construction and countdown for cold propellant models. They are more complex. And they do not boast the rip-snorting, cloud-busting performance of the solid propellant model rocket motors. But they are just as safe to operate if instructions are followed, and they can be flown in those unenlightened localities where solid propellant model rockets are still considered to be unlawful. They also provide model rocketeers with interesting alternatives to the solid propellant motors for certain applications.

Ignition and Launching

To fly a model rocket properly and safely, you must have an electric ignition system and a launching pad.

Today all model rocket manufacturers sell one or more electric ignition systems and launching pads either in ready-to-use or kit form. I highly recommend that you purchase these products, although you may be perfectly capable of making them yourself. The manufactured units contain all of the necessary and desirable safety and operational features. They may actually cost less than purchasing all the parts separately.

An ignition system and pad are what is known in professional rocketry as ground support equipment, or simply GSE. They are also called capital equipment items because, unlike a model rocket motor, they are used over and over again for many model rocket launchings. So, even though good equipment can cost you ten dollars or more, relax in the knowledge that it is a one-time expense. You can use that ignition system and launcher for flying thousands of models over several years if you take care of them and get good equipment in the first place.

To help you understand this GSE, I'll describe how to make a simple electrical ignition system and an inexpensive launcher. However, keep in mind that you are probably better off buying commercial prefabbed ignition systems and launchers.

Electric ignition devices and systems

When they launch a space rocket at Cape Canaveral, they do it electrically by pushing a button. They do not light fuses and run away.

Model rockets are also ignited by electrical means. Electrical ignition is simple, reliable, and safe. It is not only illegal to attempt to ignite a model rocket with a fuse in some states, but also very dangerous. You must not do it!

So that you will completely understand why I am so vehement about this point, here are some of the reasons why you must not use fuses:
1. Fuse ignition is not reliable. Some types of fuses will not fit into the nozzle of a model rocket motor and, therefore, will not ignite the motor. It is possible with a fuse to have a hangfire, to borrow an old gunnery term. In this case the remnants of the fuse may smolder for more than thirty minutes up in the nozzle where you cannot see them; you don't know when, how, or if ignition is going to happen. And you don't dare go to the model to find out!
2. Fuses cannot be timed. Fuse burning rates are not reliable. The fuse could flare up in a fraction of a second, and the model could take off in your face. Never, never, *never* trust the burning rate of a fuse—not when your safety and that of other people might depend on it.
3. Fuse ignition gives you absolutely no control over the launching. You cannot stop the lift-off in the last split second if you have to—not with a sputtering fuse out on the launcher. I have actually seen the following events take place during the last five seconds of a model rocket countdown: (a) a low-flying airplane appeared over the crest of a hill and flew directly over the launch site at an altitude of about a hundred feet; (b) somebody got excited and ran out to the launch pad, completely ignoring the shouts and screams of the Range Safety Officer; (c) a strong gust of wind blew the launcher over so that it was pointing toward us. Because we were flying by the rules and using electric ignition, the countdowns were stopped at once in safety. They could not have been stopped if a fuse had been smoldering and sputtering out on the launcher.
4. With fuse ignition, glowing hot remnants of the fuse are ejected from the motor nozzle upon ignition. These could fall into dry grass or other flammables around the launcher, starting a fire. Fuses have started nearly all grass fires reported to have been caused by model rockets. In several cases sputtering fuses have fallen out of the motor nozzle, landing in dry grass and starting a blaze. In general, fire marshals are now friendly to model rocketry, but not to fuses.
5. At the Cape they do not feel it is safe to ignite a space shuttle with fuses. When the professional rocketeers go back to lighting fuses

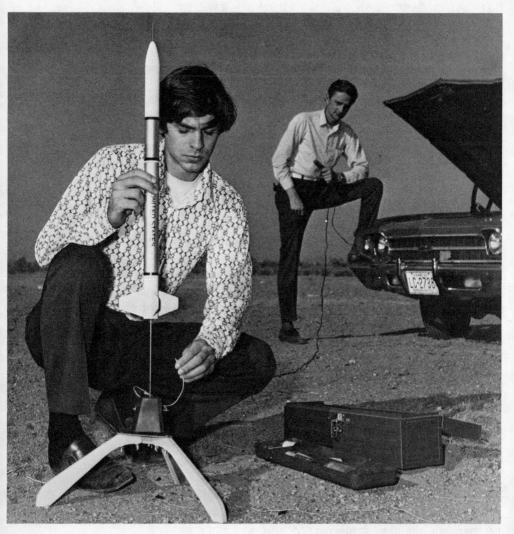

Figure 6-1: A model rocket is always launched by controlled electrical means using a launch pad and electrical ignition system.

with matches and running for the blockhouse, we model rocketeers *might* reconsider—maybe.

6. Fuses are for fireworks, not model rockets!

Simple, reliable, and safe, electric ignition gives you the opportunity to stand at a safe distance from your model and send it on its way with a professional countdown that gives you complete control over the exact moment of launching. You can stop the countdown up to the instant your finger comes down on the launch button. If there is a misfire—and who hasn't had one, even with the best preparation—you can disarm the electrical circuit with better than 99% confidence that the model isn't going to lift off.

Every model rocket motor sold today comes with an electrical igniter and with explicit, complete instructions on how to use it properly. I will discuss electric ignition in general here, but you must always read and follow what each manufacturer says about the electric ignition of his model rocket motors. Not all model rocket motors can be ignited using the methods discussed here. Often special igniters and methods must be used to ensure fast, reliable motor starts.

Nearly all electric igniters are based on the principle of the hot wire igniter. This is a short piece of very fine nichrome wire. It works just the way the electric heater element in a toaster does. When an electric current is passed through the wire, its electrical resistance, or inability to pass the electric current, causes it to get hot. A hot wire igniter raises the temperature of the adjacent solid propellant surface to the temperature necessary for ignition. The ignition temperature for most model rocket motors is in excess of 550° F. Once the propellant surface is raised to this temperature and starts the process of combustion, the combustion process continues in a sort of chemical chain reaction until all propellant is consumed.

A typical hot wire nichrome igniter is shown in Figure 6-2. For motors up to and including NAR Type C, a 2-inch length of nichrome wire will be sufficient. Double it over, and insert the folded middle part up the motor nozzle until it comes into contact with the propellant and won't go any farther; hold it in place with a piece of flameproof wadding jammed into the nozzle, and you will have your igniter all ready to go with two tiny ends sticking out.

Don't worry about the wire or the wadding. Once the motor starts, both will come out rapidly. In fact, a nichrome wire igniter is good for only one shot.

Nichrome wire is made from an alloy of nickel and chromium. Used for electrical heating elements, it is designed to have high

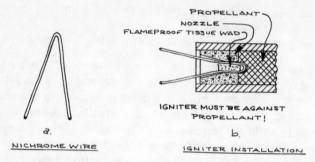

a.

NICHROME WIRE

b.

IGNITER INSTALLATION

Figure 6-2: Drawing of a typical nichrome wire igniter and its installation in a model rocket motor.

electrical resistance and to withstand the high temperatures generated when electrical current passes through it. The nichrome wire used in most model rocket igniters is, technically, AWG No. 30 or AWG No. 32, having a wire diameter of 0.008 inch to 0.012 inch. The thicker or heavier the wire, the lower its electrical resistance.

The amount of electrical current that will flow through any conductor is determined by Ohm's Law, which is written:

$$I = E/R$$

where E = voltage, R = the electrical resistance in ohms, and I = the current in amperes that will flow. The higher the voltage, the more current will flow. The higher the resistance, the less current will flow.

It takes electrical power to cause an igniter to heat up. In an electrical circuit power is determined by the equation:

$$P = I^2R$$

where I = current in amperes and R = resistance in ohms. P = power in watts. This equation tells us that if you increase the resistance, the power that is dissipated in a circuit, and therefore the heat that is developed, is halved. But if you double the current, the power dissipated in the circuit goes up four times!

Together, these two equations indicate several things. First, if the voltage is too low, not enough current will pass through the igniter to generate enough power to raise the temperature to the point where the solid propellant will ignite. However, if you double the voltage, twice the amount of current will flow, and this will increase the power and the heat by a factor of four.

In a typical nichrome igniter the resistance is about 1 ohm. If you use a 6-volt battery and a good wiring system that itself has 1 ohm of resistance or less, there will be about 16 watts of power in the igniter. This is just about the minimum required to get an igniter sufficiently hot. However, if you use a 12-volt battery, it will have to deliver only a quarter of the amount of current to achieve the same igniter temperature.

Using nichrome wire 0.010 inch in diameter, I made some tests with electrical measuring instruments hooked into the ignition circuit. I found that 5 volts is just barely enough to get the igniter warm; 6 volts is minimum; 12 volts does the job nicely; 18 volts or greater burned the nichrome wire in half before it got hot all over.

Never plug an igniter into the 115-volt AC house circuit. This is nearly a direct short circuit for the 115-volt line, because the theoretical igniter current will be about 90 amperes, which is more than your circuit breaker is designed for. If it doesn't trip the circuit breaker right away, the igniter itself will simply snap in two in reaction to this mighty surge of current. You may also damage your ignition system beyond repair because it is not designed to handle this sort of voltage and current.

A properly designed and constructed electrical ignition system must have very low resistance so that it doesn't get hot instead of the igniter! Building a good ignition system is not difficult. Even if you buy and build a kit, however, you should build it correctly.

The parts of an electrical ignition system are a launch controller, a battery, and connecting wires.

Basically, the launch controller is a device that makes sure no electrical current gets from the battery to the igniter before you want it to do so; then, when you complete its program, it delivers enough current to the igniter to ignite the model rocket motor. Its basic part is a spring-loaded push-button switch that will remain in the off, or open, position and will automatically return to the off position when released. An ordinary electric light switch or a knife switch will not make this automatic return, so it is extremely hazardous to build your own launch controller using such switches. Sooner or later you will forget to return the switch to the safe off position after launching. Then, the next time you put a model on the launch pad and hook up the igniter, the model will take off in your face. Having this happen to you once is more than enough, believe me.

All commercial launch controllers have the proper spring-loaded, normally open push-button launching switch. If you are building your own launch controller, you can use an ordinary doorbell button, as shown in Figure 6-3.

Safety rules also require a safety key, which, when removed from the launch controller, opens the electrical circuit at yet another point. It provides a double precaution, something that engineers call "redundancy." When the safety key is out of the controller and in your pocket, there is no way for any electricity to flow out to the igniter in the model. Nobody is able to launch that model except you. Nobody can get cute and try a practical joke, pushing the launch button just when you've finished the igniter hookup. This is why you must have a safety key and why you must *keep it with you*. Don't leave it in the launch controller, or tied to the controller. Keep it in your hand, in your pocket, or on a string around your wrist or neck. I emphasize/this because it is vital to your safety, and you should not forget it for a single instant when you are launching. With the safety key in your possession, you and *only you* can launch that model. You can work around the model on the launch pad knowing that nobody is going to play games with you and launch the model while your fingers are under it. If you put a big bright-colored tag or flag on the

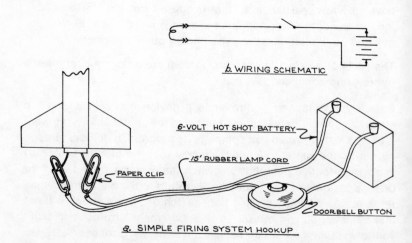

Figure 6-3: Drawing showing the construction and hookup of an inexpensive electrical ignition system.

safety key, you will be able to see if you have accidentally left it in the launch controller.

Once the safety key has been inserted into the launch controller, the controller is said to be armed. This means that it is ready to launch the model when you push the switch.

A wiring diagram for a typical electric ignition system is shown in Figure 6-4. You can see how the safety key opens the circuit. Notice that there is also a continuity light wired across the ignition switch. This is a small light bulb that lights up when you insert the safety key; it tells you that you have a completed circuit. The continuity light allows a very small amount of current to pass through the circuit and igniter, but not enough to get the igniter warm. It is a limiting electrical resistance. When the ignition switch is pushed and closed, the light bulb is shorted out, allowing the full battery current to flow through the igniter.

The continuity check light will *not* tell you that you do not have a short circuit between the clips or between the clips and any metal launch pad parts. That you have to check for yourself when hooking up the igniter. It *will* tell you, by failing to light up, that you have dirty clips, and the remedy for this is covered in the troubleshooting section later.

Except for use with very special igniters, the wire used to connect the battery, controller, and igniter—called the connecting wires or

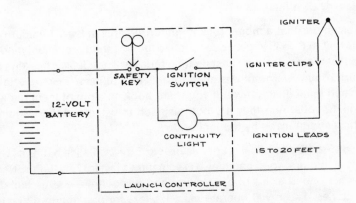

Figure 6-4: Electrical wiring diagram of a typical electrical launch controller showing safety key and continuity check light.

firing leads—should be heavy-duty stuff. Don't wire up your system with doorbell wire, although this is commonly available in most hardware stores. It is too small (AWG No. 22, usually) for use in the heavy current model rocket ignition systems and has far too high an electrical resistance. Rubber-covered AWG No. 18-2 lamp cord works very well and comes with two conductors that are insulated. This rubber-covered lamp cord, commonly called zip cord or POSJ, is also available in hardware stores. Or you can cut up an extension cord to get the wire.

You will need something to hook the firing leads to the igniter. Commercial systems come with tiny spring clips called micro-clips. These are also available at most radio and electrical stores. Don't use the big, heavy alligator clips because their weight will usually pull the igniter right out of a motor. You can use paper clips to make the hookup, as shown in Figure 6-3.

When constructing your electrical ignition system, build it to last! Solder all connections, if possible. Don't just wrap wires together; solder them or clamp them under a screw head. Remember, your electrical ignition system will pass enough current to operate a small electric oven; it must be able to pass enough current to light a 500-watt light bulb without trouble. Your personal safety depends on its being built correctly.

Batteries give beginning model rocketeers more trouble than anything else. But such trouble won't happen to you if you will take the time to read and heed the following section. I have rarely had a launching battery fail me; in those few instances when one did, it was my own fault.

Remember that a model rocket igniter requires at least 1 ampere of current at 6 volts to get hot enough to ignite a solid propellant model rocket motor. Ordinary flashlight batteries will not deliver this kind of current and were not designed for this sort of use. They are totally ruined immediately, even if you try to hook up eight of them to give you 12 volts. And they are not designed to withstand the seeming short circuit made by the igniter.

Four Size D photoflash batteries or four Size D alkaline batteries will make igniters work, and they will sometimes last for fifty flights. Next up the line in terms of battery size and power are the large lantern batteries. These will operate most igniters for one hundred flights, maybe more. Some model rocketeers buy the larger Hot Shot batteries that will last for six months to a year of regular weekend flying.

Table 6
Recommended Battery Chart

Type	Volts	Eveready	Bright Star	Burgess	Mallory	Marathon	Ray-O-Vac	RCA	Sears	Ward	Wizard
D Cell Energizer (8 required)	1.5	E95BP	7520	AL-2	Mn-1300	122	—	VS1336	4653	—	—
D Cell Photoflash (8 required)	1.5	850	10P	220	M-13P	124	210LP	VS736	—	3228	—
Radar-Lite	6	731	158	TW-1	M-918	896	918	VS317	4707	8MW	7D8918
Hot Shot	6	1461	146	S461	M-907	640	—	VS039	4668	7MW	7D8907
Lantern	7.5	715	155	4F5H	903	903	903	VS139	—	—	—
Lantern	9	716	164	4F6H	M-904	904	904	VS140	—	—	—
Radar-Lite	12	732	—	TW-2	—	732	926	VS342	—	—	—
Hot Shot	12	1463	187	2G8H	—	642	—	—	2335	—	—

Several manufacturers make the same type of battery as shown above. All those making a given type are listed.

All of these batteries are dry cells or modifications of the LeClanche cell. Table 6 shows the most commonly used dry cells for occasional weekend flying. The number of flights you can expect from them depends on their age, how long they sat on the store shelf before you got them, and the temperature on the day you are flying. If these batteries get cold, they will not put out as much current. A dry cell battery that works fine on a summer day may fail to ignite a motor in the cold of winter. Below 45° F. dry cells get cold and sluggish and are not very happy about igniting model rockets. But if you are on a tight budget or are just too lazy to get a better battery, dry cells will work for you if you keep them warm in a car or in your pocket.

The ultimate in compact model rocket ignition batteries are Size D sintered-plate nickel-cadmium rechargeable batteries made by Gould-National, Union Carbide, Eveready, and Burgess. These little jewels will throw a current of 50 amperes through a dead short for 90 seconds before they go dead—and you can recharge them so that they will do it again. Ten of them in series are equivalent to a full-sized car battery; in fact, I have used them to start my car on a cold morning when the car battery was dead. Their big problem is their high cost—about five dollars apiece or fifty dollars for a set of ten of them to make 12 volts! I managed to promote a set of these beauties from a battery manufacturer many years ago for evaluation purposes, and evaluate them I did. They lasted for about seven years and five hundred rechargings before they finally quit for good—and I wept buckets over their dead, expensive little carcasses.

Auto batteries are probably the finest of all batteries for model rocket ignition work. You can connect the battery clips of your electric ignition system right across the battery terminals without removing the battery from the car. A 12-volt auto battery is one hot ignition source; there is literally no delay between the time you push the launch button and the time the motor ignites.

One model rocket manufacturer put out an electrical system with an adapter that plugged into a car cigarette lighter, allowing the use of the car battery without any fuss whatever. You can buy an adapter like this in most auto parts stores.

The big problem with an auto battery, however, is its weight. There comes a time when you want to have it in the middle of a field where you can't drive a car or in a schoolyard where cars aren't allowed. Since most car batteries weigh about 60 pounds, they are somewhat heavy for 97-pound weaklings to remove and carry around.

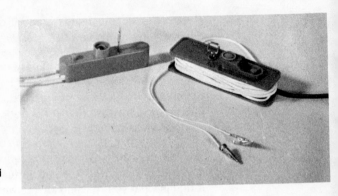

Figure 6-5: These commercial electrical controllers with their built-in safety keys, push-button launch switches, and continuity check lights are made by Estes Industries, Inc., (left) and Centuri Engineering Company (right).

The ultimate solution is a motorcycle battery like those used on a Honda, Yamaha, Kawasaki, etc. These little batteries are miniature auto batteries. Readily available at most motorcycle stores, they cost between ten and fifteen dollars. They weigh about 5 pounds, are small and easy to carry, and will launch over a thousand model rockets before needing a recharge, which is plenty for a weekend of flying.

An auto or motorcycle battery is technically known as a lead-acid battery. Properly cared for, it will last for years. You must keep it charged up, and you must keep the electrolyte level at the specified point by adding *only* distilled water. Tap water usually contains chlorine and other chemicals that will corrode lead-acid cells. As a matter of fact, you can make any auto battery outlive its warranty by adding only distilled water to the initial battery electrolyte; for example, a cheapie twelve-month-warranty auto battery will last over five years if only distilled water is added to it. Distilled water is available in most supermarkets for use in electric steam irons.

When you buy a lead-acid battery these days, it is dry-charged, meaning that it does not have any electrolyte, or battery acid, in it. Normally a container of electrolyte is included in the purchase price of the battery, but sometimes you may have to buy it separately at an auto parts store or the place where you bought the battery. Battery electrolyte is a mixture of sulphuric acid and distilled water adjusted so that its specific gravity is about 1.280. It is put in the battery only once; thereafter only distilled water is added, as needed, to maintain the proper electrolyte level.

When you pour the electrolyte into the battery, be careful. It will make short work of your clothes if you spill any on them. If you get electrolyte on your hands, wash it off at once with running water. Flush down any spills with water.

You will need to obtain a battery charger if you have a lead-acid battery—unless you want to take it to the local service station weekly or monthly for a *slow charge*; don't let the mechanic put a fast charge on it because that is not good for the life of the battery—and you want that battery to last!

You can buy an inexpensive battery charger for auto batteries, but check to see that it isn't too powerful to charge a motorcycle battery, if that is what you're using as a power source. Normal charging rate of a lead-acid battery is one-tenth the rating in ampere-hours. That is, if you have a 40-ampere-hour battery, charge it for 16 hours at 4 amperes.

You may already have a battery charger that will handle most motorcycle batteries, and you probably don't know it. If you have an HO- or N-gauge model railroad or a slot car racing set, you can use the power pack as a battery charger. This pack converts 115-volt AC house current to the 12 to 18 volts DC needed to run trains or slot cars. Regardless of whether you have a throttle pack or a slot car power pack, you will need an ammeter to measure the charging current. Most motorcycle batteries will charge very well at 0.4 ampere for 16 to 18 hours, and this is well within the current range of most power packs. You won't use the full capacity of a 0-1 ampere DC ammeter, but this is what you should have. Such an ammeter can be obtained from a radio supply shop, a hobby store that handles radio-controlled model airplanes, or a mail-order radio supply house such as Allied Radio Corporation, 100 North Western Avenue, Chicago, Illinois 60680.

If you want to order your meter from Allied, you can get an inexpensive Eico Model NF-2C for about $2.00 (Allied stock number 68 U 972) or a somewhat better one, the Eico Model RF-2-1/4C, for about $2.50 (Allied stock number 68 U 025).

If you use an HO throttle pack with its adjustable throttle as a battery charger, you will need nothing more than the ammeter, some hookup wire, and some clips to attach the hookup wire to the battery terminals. Wire the circuit as shown in Figure 6-6. *Watch polarity!* You can burn out the meter or ruin the battery if you do not hook up with the correct polarity. You can test for polarity in this way. When you finally turn on the switch and turn up the throttle control a little,

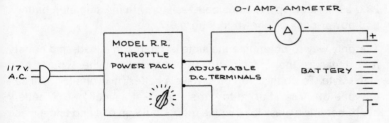

Figure 6-6: Wiring diagram for charging lead-acid battery with an HO train throttle pack.

the polarity is wrong and the battery is hooked up backward if the ammeter needle tries very hard to go off-scale to the left or below the zero mark. The remedy: Flip the direction switch on the throttle pack or reverse the terminal connections at the battery.

When the polarity is correct, turn the throttle control up until the ammeter reads the recommended charging current for your battery. Check the reading every hour or so and adjust the throttle control to adjust the amperage. As the battery charges, the charging current will decrease, and you will have to increase the throttle setting to maintain a constant charging current.

If your battery charger is a slot car power pack without the adjustable throttle of the HO train pack, you will have to add an adjustable resistor, or rheostat, so that you can adjust the current coming from the power pack. In other words, you will need something to take the place of the throttle control of the train throttle pack. This replacement is an adjustable resistor of 30 to 50 ohms resistance. Suitable adjustable resistors, called "pots" for "potentiometers" in electronic parlance, include: IRC-CTS 4-watt, 30-ohm, wire-wound control for $1.32 (Allied stock number 30 U 216 C) or the Mallory Type VW-30 5-watt, 30-ohm, wire-wound control for $1.39 (Allied stock number

Figure 6-7: Wiring diagram for charging lead-acid battery with a slot car power pack and rheostat, or adjustable resistor.

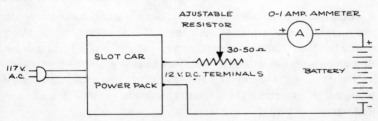

28 U 097). Hook up the adjustable resistor with the ammeter, battery, and power pack as shown in Figure 6-7.

The only way to determine the state of charge in a lead-acid battery is to measure the specific gravity or strength of the sulfuric acid electrolyte. To do this, you must have a battery hydrometer. You can buy one of these at an auto parts store. Get a small one. A battery hydrometer is a glass tube with a rubber bulb on one end and a piece of small tubing on the other. There are four little colored balls or a float inside the glass tube. Stick the tubing down into the electrolyte of one cell of the battery, draw enough into the glass tube to float the balls or the float, and check the number of balls floating or the specific gravity reading on the float. This will tell you the state of charge of that battery cell. You must check all cells. One or more of them could be slightly discharged.

Even if you have charged the battery for the recommended number of hours at the recommended charging current, you may have to put a little more into it to get all cells up to full charge. It becomes increasingly difficult to bring a cell up to the final charged level as it finishes off its charge.

A lead-acid battery produces hydrogen gas when it is charging. So do not charge one in an enclosed place. Choose a place with ventilation. A garage will do because most garages are not airtight.

Keep your lead-acid battery clean. Remove any corrosion around the caps or terminals by washing with water. If the corrosion is bad, wash with a solution of baking soda—but don't let any of this liquid get into the electrolyte!

Keep your lead-acid battery charged. A battery that just sits on the shelf will undergo self-discharge after a month or so. Don't let the cells discharge themselves, and don't store the battery in a dis-charged condition. A discharged battery allows lead sulfate to form in the cells. Lead sulfate is insoluble in electrolyte and does not conduct electricity. As a result, lead sulfate eventually coats the battery plates, and the battery becomes sulfated. Kiss it good-bye.

It is possible to desulfate a battery that has gone only partway toward this final resting place for all lead-acid cells. You must charge it very slowly at the recommended charge rate for a long, long time. Sometimes it takes weeks to desulfate a battery. Often batteries are so far gone that they can't be ressurected. You must do a lot of charging to overcome the neglect that leads to sulfation, so keep that battery charged!

You can ensure a full charge on your battery by putting it on trickle charge during the flying season when you don't have it in the field. A trickle charge is a very low charge rate. The normal trickle charge rate for a battery is 10% of the recommended full-charge rate. In other words, if your battery requires 0.4 ampere at a full-charge rate, the trickle charge rate will be 10% of that, or 0.04 ampere, just about the minimum that you can read on a 0-1 amp meter. The battery should not give off bubbles of gas while it is trickle-charging; if it does, reduce the current until the gassing stops.

Although lead-acid cells require more care and attention than dry cells, they are more reliable for model rocket ignition. They will also last a very long time and, in the long run, save you lots of money.

Rechargeable nickel-cadmium batteries can be charged in the same way as lead-acid batteries. However, they are sealed, and you cannot determine the state of charge by measuring the specific gravity of their electrolyte. The best bet with Ni-Cd batteries is to keep them on a 10% trickle charge when they are not in use.

Launch pads

Good electric ignition is only one part of getting a model rocket off the ground for a successful flight. You must also have a launcher, or launch pad.

A model rocket launcher is a device that allows the model rocket to gain flying speed before becoming airborne. If you were to set a model rocket on the ground on its fins with its body pointed straight up, it probably would not fly straight. Sitting on the ground, the model has no air flowing over its fins to maintain its stability and keep it going in the direction it's pointed. About a quarter of a second is required for most model rockets to build up thrust and accelerate to a speed of about 30 miles per hour where the fins can properly act to stabilize the model.

Without a launcher a model rocket is likely to topple over during the first split second of flight. If this happens, it will be all over the lot. And you can't outrun it. For safety a launcher must *always* be used.

A rod launcher is the simplest and most commonly used type of launcher. Various styles are available at reasonable prices from

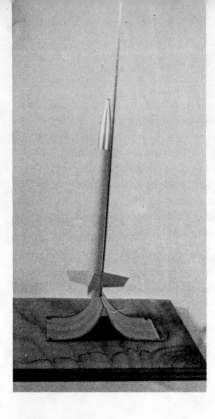

Figure 6-8: A simple rod-type launch pad with plywood base, launch rod, and jet deflector cut from a tin can.

model rocket manufacturers. Most of them have the following basic features.

A rod launcher is nothing more than a piece of 1/8-inch diameter hard steel wire at least 36 inches long. It is held in a vertical or near-vertical position by the launcher base. The 1/8-inch by 36-inch rod is the NAR Standard Launch Rod.

Most launch pad kits come with a two-piece launch rod. This allows it to be put in a shorter package. A two-piece rod is also handy to carry around, but you are always in danger of losing or forgetting half of it. Half a launch rod will keep your birds out of the air because you should never attempt to launch a model rocket with less than a 36-inch rod.

If you want to make a two-piece rod into a one-piece unit, stick a piece of solder 1/8 inch long into the hole of one of the rods. Stick the other rod into the hole and hold the assembly over a match or candle until the solder melts to hold the rods together. Or you can go down to the hobby store and buy a 36-inch length of 1/8-inch diameter music wire. This is the very hard steel wire used to make landing gear of model airplanes. The longer launch rod is clumsier to carry around, but you are always certain of having a full-length rod with you for flying.

A simple but effective launch pad base is nothing more than a piece of 3/4-inch plywood about 1 foot square. Drill a 1/8-inch diameter hole in the middle of the board and at right angles to the board. Tap the launch rod down into the hole until it wedges into place.

More sophisticated launch pads are usually made of a central piece and three adjustable and/or removable legs forming a tripod. A good launch pad base has legs spread widely apart for stability; when the launcher sits on the ground, it should be very difficult to tip over. In addition, the three legs should make it stable on uneven ground. The launch pad base must be solid and stable so that a little gust of wind doesn't tip it over.

Figure 6-9: The commercially made launch pad at left is from Estes Industries, Inc. The commercial unit at right from Centuri Engineering Company includes battery and electrical system as part of the unit assembly.

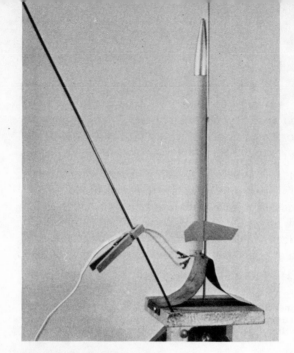

Figure 6-10: An umbilical tower and clothespin support the weight of the electrical leads.

A launch pad should also have a tilting mechanism so that the launch rod can be tipped slightly away from the vertical position. This allows you to put a bit of tilt in the launch rod to compensate for wind drift and weathercocking, which we'll talk about later. The safety rules say that you must always launch within 30 degrees of the vertical, however, and all commercial launch pads are designed so that this limit cannot be easily exceeded. Tilting is usually achieved by adjusting one or more of the launch pad legs or by a tilt-leg adjustment on only one leg. Always make sure that the tilting mechanism is tightened so that the model cannot tip the launch rod by its weight. Also make sure that the launch pad cannot fall over easily with the model in place for launch. If necessary, put one or two rocks on the base or legs to hold the launcher in place.

Some commercial launch pads are available with the ignition battery built into the launch pad base for weight and stability. This dual-purpose design makes a very stable launch pad.

A launch pad should always have a jet deflector to prevent the motor exhaust jet from striking the launcher or the ground. The deflector must be made of steel and is usually a steel stamping. Although a flat metal plate can be used as a jet deflector, it can turn the exhaust jet back on the model in the split second before the model lifts off; this can discolor or burn the fins and tail. A better jet deflector is a bent or angled piece of steel that turns the jet at right angles and streams it away from the model, pad, and ground. This is important, for if the launch site is covered with dry grass, the deflector can prevent your model from starting a grass fire.

A jet deflector can be cut from a tin can. Always use a steel can. Deflectors made from aluminum or cut from aluminum cans will not stand up under the temperature of the exhaust jet. A hole will be burned through them after only a few flights. A steel jet deflector will last for a long time.

Another launcher feature that is very helpful is an umbilical tower. This is a rod or dowel standing a few inches to one side of the launch rod; it has a clip, clothespin, or a piece of tape for holding and supporting the weight of the firing leads and clips. Some model rocket igniters have a tendency to pull out of the nozzle when they start to get hot; or the tape disc that holds the igniter in place can come loose and let the igniter fall out. Many, many misfires are caused when the relatively massive weight of the firing leads drag down on the igniter, pulling it out of the motor nozzle before it has a chance to get the motor started. An umbilical tower avoids these problems by supporting the major weight of the firing leads, leaving only a few inches of firing leads for the igniter to support.

If I am flying from somebody else's pad that does not have an umbilical tower, I will tape the firing leads to the bottom part of the launch rod or to one of the launch pad legs to take the weight of the firing leads off the igniter.

An umbilical tower is also a must when flying front-motored boost-gliders, as we will see in a later chapter.

To prevent accidental eye injury, you should also make a rod cap for your launcher. This can be nothing more than an old motor casing painted a bright fluorescent color and placed over the top of the launch rod when the launcher is not being used on the flying field. It alerts you to the location of the rod end so that you don't run into it. You might also attach a plastic streamer to a spring clothespin and clamp the clothespin to the top of the launch rod so that the fluttering streamer alerts you. You can further protect your eyes by wearing sunglasses while flying.

Be sure to remove the rod cap or streamer from the top of the rod before launching your model. One young rocketeer put his model on the rod and attached a clothespin to the top of the rod because he wasn't going to launch right away. Later he forgot to remove it. The rocket took it off when it left the rod. This caused a rather spectacular flight in some wildly unpredictable directions because the clothespin slowed the model for a split second just as it was leaving the launch rod.

If you fly from a grassy field, set your launch pad up on a tarpaulin or blanket. This will not only keep the knees of your pants dry and clean when you are hooking up, but will also prevent grass fires. The tarp will catch any glowing pieces of igniter wire or wadding that might be ejected from the motor nozzle. It's a precaution that is worth the trouble.

There are several auxiliary items that you should have tied to your launch pad with a string or at the launch pad site. One is a roll of paper masking tape 3/4 inch wide. There may be times when you need to tape things up at the pad. For example, some models may require support on the rod to keep their tails from resting on the launch pad base. Tape can be used as shown in Figure 6-11 to hold a model up on the rod. But don't forget to give the model all the rod length possible; it will be more stable in flight.

Figure 6-11: A model can be held up off the launch pad base by a piece of paper tape wrapped around the launch rod as shown.

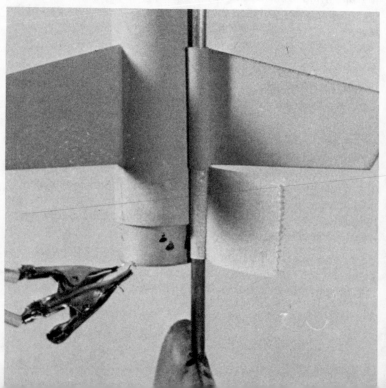

At the launch pad you should also have a small piece of sandpaper or an emery board such as that used to file fingernails. This is known as a J.I.C. file. J.I.C. stands for "Just In Case," and the file is used to remove the residue that forms on the firing lead clips, thus exposing a clean metal surface that will make a good electrical contact with the motor igniter.

I also keep a few spring clothespins in my launcher box. They are handy for supporting a model on the rod, for clamping firing leads to launch pad parts, and for handling the hundred other little clamp jobs that arise at the launch pad.

I'll not forget the time a news camera crew from a major TV network came out to film a flying session of our model rocketry club. They had just returned from Cape Canaveral and a major Apollo lunar mission. We put a big 1:100 scale Saturn-V on the pad, and I went underneath it to hook up. Cameras ground away with microphones catching every breathless prelaunch word. Then I blandly turned to a fellow rocketeer and said in a matter-of-fact tone, "Hand me that clothespin over there." The TV crew, used to the intensity and drama of a full-scale Saturn launch, was completely thrown.

Earlier I mentioned that a model rocket should never be launched at an angle of more than 30 degrees from the vertical. Most commercial launch pads have stops on their tilt mechanisms to prevent rocketeers from exceeding this 30-degree tilt limit. Why this limit? One word, and you have heard it before: *safety*.

It can all be traced to an Italian Renaissance artillerist named Tartaglia, who discovered and wrote down what any artillery soldier still calls Tartaglia's Laws of gunnery. Tartaglia's Laws are as true for

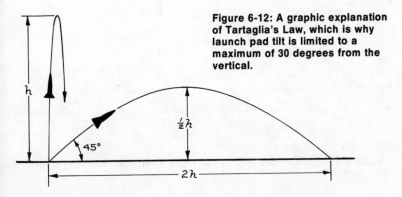

Figure 6-12: A graphic explanation of Tartaglia's Law, which is why launch pad tilt is limited to a maximum of 30 degrees from the vertical.

model rockets as for guns. Written in modern model rocket terms, they state:

1. A model rocket that will go 1,000 feet straight up will travel a horizontal distance of 2,000 feet if launched at an angle of 45 degrees.

2. When launched at an angle of 45 degrees, a model that would go 1,000 feet vertically will ascend to a peak altitude of only 500 feet during its arcing flight.

3. The path followed by the model rocket when launched at any angle other than the vertical will describe a parabola. (For very long-range shots such as an ICBM or space vehicle, the path becomes an ellipse because the curvature of the earth must then be taken into account. See any physics book, gunnery table, or computer program for interplanetary trajectory guidance of space vehicles.)

I can report that Tartaglia's Laws are correct. I tested them in the wide-open space of the American West under careful controls; you don't have this much safe flying area, so this is one scientific experiment that you should not perform. There are many other safe experiments that you can perform with model rockets, so launch within 30 degrees of the vertical, the way the safety rules specify.

To answer a question nearly everyone asks at first: No, a model

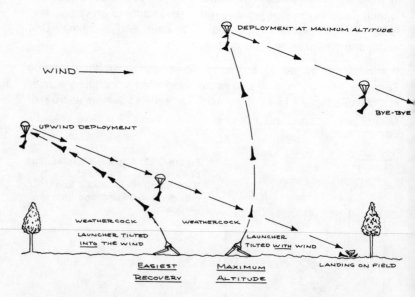

Figure 6-13: You can use launch pad tilt to make your model fly where you want it to.

rocket will not fly horizontally. It has no wings to provide lift against gravity, and its fins won't do the trick. When launched horizontally, a model rocket merely flops to the ground off the end of the launch rod. It then skitters around on the ground during the thrust period. This performance bends the model more than just a little bit.

The reason for permitting a launcher tilt of up to 30 degrees from the vertical is to allow the model rocketeer to compensate for wind effects on his model during flight. In a wind a model rocket will exhibit launcher tip-off just as it leaves the launch rod. The horizontally blowing wind makes the model cock its nose into the wind just like a weathervane, or weathercock. And that is what we call the phenomenon—weathercocking. The amount of weathercocking depends upon the velocity of the wind, the weight of the model rocket, the stability margin of the model, the acceleration of the model, and other factors.

Model rocketeers learn to make use of the weathercocking phenomenon. If you can't fool Mother Nature and if you can't fight her, maybe you can use her.

To gain maximum altitude during a model rocket flight in windy conditions, rocketeers tilt the launch rod *downwind*. They adjust the tilt mostly by experience, using the "wet finger" method and a knowledge of how their model will perform in a wind. When the model lifts off, the weathercocking effect makes it tilt *upwind*. As a result, the model ends up going straight up over the launch pad. However, if you try for altitude in strong winds, your model is going to drift a very long distance after the recovery device deploys, so you'd better have plenty of open field in the downwind direction.

For maximum duration flights, or when flying on small fields, you'd best tilt the launch pad *into* the wind. This will make the model weathercock even more into the wind. It then drives upwind so that the recovery device activates well over the upwind side of the field, the model drifts back over the field and launch site, and hopefully the entire affair touches down just before reaching the rocket-eating trees on the downwind side of the launch field.

Although most model rockets weighing less than 6 ounces (170 grams) can be launched safely from a 36-inch or a 1-meter launch rod, this rod length should not be used for heavier models unless high-thrust motors are used. Heavier models require more time and distance to build up stable flying speed, and you should give them all the launch rod that you can for best guidance. There is absolutely

Figure 6-14: Tower launchers such as this one at NARAM-14 are coming into increasing use. This tower can be adjusted to accommodate body tubes of various sizes.

nothing to be gained by using a short launch rod at any time. Always use the longest launch rod that you can. Models heavier than 6 ounces should be launched from rods that are 5 to 6 feet long with diameters of 3/16 inch or 1/4 inch. You may be able to find launch rods of this size at a local hardware store or a scrap metal dealer. Or you can call a local steel supply company; a quick look in the Yellow Pages of the telephone directory will guide you to the right place.

Naturally, when using the larger and longer launch rods, you must put bigger launching lugs on your models so that they will fit over the rod.

Another type of launcher seen on some model rocket ranges, especially during NAR competitions, is the tower launcher. This is a structure, often very simple in construction, in which the model slides between guide rails. There may be three or four guide rails. No launch lugs on the model are required. This reduces the aerodynamic drag of high-performance, high-efficiency competition model rockets. A typical tower launcher is shown in Figure 6-14.

There are advantages and disadvantages to a tower launcher. On the positive side, as mentioned above, a tower eliminates the need for launch lugs, which reduces the drag of the model. Also, a tower provides a very stiff launcher that does not sway or bend or whip as the model is launched. On the negative side, a tower is a complex piece of GSE that easily gets out of adjustment, especially when you carry it around in the trunk of a car. The guide rails must be carefully adjusted for each different model rocket body diameter. And some competition modelers believe that the friction of the three or four tower rails is greater than that of a single launch rod and launching lug. They can show equations that prove to them that this cuts down the maximum altitude of the model.

Tower launchers are built from wood, plastic, or metal. An increasing number of designs have appeared over the past decade.

The third type of launcher is the rail launcher. It has also seen increasing use in model rocketry, especially with large and heavy models. It is often called a C-rail launcher, because in cross section it looks like the letter C; see Figure 6-15. A model designed

Figure 6-15: A C-rail launcher engages a standard launch lug on a standoff as shown and is stiffer than a regular rod.

to use this sort of C-rail launcher can also be used on a standard rod because of its launching lug. Another type of rail launcher is the split-lug type shown in Figure 6-16. Again, a model designed for use with a split-lug launcher can also be launched from a standard rod. The primary reason for using a rail launcher is to take advantage of the stiffness and lack of whipping of this stronger design.

A simple trough-type launcher such as once used to fire skyrockets is not safe and should not be used for model rockets.

Remember that during launch a model rocket must be restrained from moving in any direction except the flight direction. The launcher performs this task. It is a model rocket's guidance system.

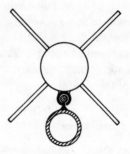

Figure 6-16: Cross section of a split-lug launcher and model.

How High Will It Go?

Once a model rocket leaves the launcher, it is a free body in space, even though it is still surrounded by the earth's atmosphere. It has been projected beyond the earth's surface, and its actions as a free body in space cannot be duplicated easily while it is on the ground. But we can account for the effects of the earth's atmosphere, and if we subtract these effects, we can study the motion of the airborne model just as if it were in outer space. We can discover where it will go, how far it will go, and how fast it will go.

It is possible to "fly a model rocket on paper." All you need are the elementary tools of simple arithmetic—addition, subtraction, multiplication, and division—a pencil and paper—and an eraser if you are prone to making mistakes. With just these you can find out in advance of flying your model exactly how it will perform. Honestly, you don't have to be a genius to do it. And it is very exciting to work out the numbers and then have the model perform the way the numbers said it would.

Although a model rocket flight has three basic phases—powered flight, coasting flight, and recovery—we are going to discuss only the first two phases here. Recovery will be treated separately in another chapter.

Three basic forces act upon a model rocket in flight. You can think of a force as the application of energy to the model in such a way that the flight path is changed. As shown in Figure 7-1, these three forces are:

1. *Thrust* from the model rocket motor that acts on the back of the

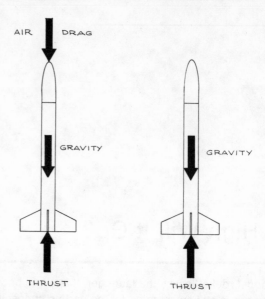

Figure 7-1: Diagrammatic drawing showing the forces on a model rocket in flight.

Figure 7-2: Diagrammatic drawing showing the forces on a model rocket in drag-free flight.

model and makes it accelerate, or gain speed.

2. *Gravity* that tries to slow the model and pull it back to earth again. In vertical flight gravity acts exactly in the opposite direction to thrust.

3. *Aerodynamic drag* caused by the model ramming its way through the air; this drag force also acts to slow the model.

During powered flight all three forces act upon the model. During coasting flight thrust is zero. Therefore, only gravity and aerodynamic drag act on the model.

To begin with, we are going to assume that there is no aerodynamic drag on the model. In other words, we are going to ignore the effects of the earth's atmosphere and pretend that our drag-free model is acted upon only by thrust and gravity, as shown in Figure 7-3. By figuring the flight of the model as though it were flying in the vacuum of space, we will see how very important proper streamlining is, because this drag-free flight will be quite different from flight with drag. You will see what a tremendous amount of aerodynamic drag is exerted on a model in flight.

Although model rocketry is conducted in the metric system, we will use the more familiar English system here, because many of you may

Figure 7-3: This is the first photograph of a model rocket in flight, taken in 1957 at a shutter speed of 1/1000 of a second. The speed of the model is evident.

not yet be really familiar with metrics. Most Americans are still more at home with feet, inches, pounds, and ounces. It is simple enough to convert from metric to English and back again. Just refer to the conversion chart given in Table 3 in Chapter 3.

To better understand what happens to a model rocket in flight, let's briefly review part of the basics about the motion of bodies in space.

When a body moves from Point A to Point B, it covers the Distance S between the two points. We are considering here only simple motion in a straight line in one dimension; add other dimensions later if you want to. Since the body cannot go from Point A to Point B in zero time, it takes a finite period of Time T to cover the Distance S. If $S = 1$ foot and $T = 1$ second, the body is said to be moving with a *velocity* of 1 foot per second. Velocity is then defined as the distance traveled divided by the time required for travel. During the next interval of time, T_2, the body will cover another distance equal to S if its velocity is constant.

If the velocity of the body is not constant but changes, the body is said to be *accelerating*. It changes velocity at a given rate. A car accelerates when it goes from 0 to 60 miles per hour in 6 seconds. If at the beginning of the interval of time our hypothetical body above is moving at a velocity of 1 foot per second, and at the end of that interval of time it is moving at a velocity of 2 feet per second, it has changed its velocity by 1 foot per second during that time interval. If the time interval is 1 second, the body has experienced an accelera-tion of 1 foot per second per second, or 1 ft/sec/sec, or 1 ft/sec^2.

In a like manner, going one step further, if the acceleration changes by 1 ft/sec^2 during a 1-second interval of time, the body has experienced a rate of change of acceleration, or surge, of 1 ft/sec^3. But don't worry about it.

In 1687 the theologian and astrologer/physicist Sir Isaac Newton published a document entitled *Philosophiae Naturalis Principia Mathematica*. In this classic paper, known simply as the *Principia*, Newton revealed his famous three Laws of Motion. Simply stated, they are:

I. Law of Inertia: A body at rest will remain at rest or a body in motion will remain in motion in a straight line with constant velocity unless acted upon by an exterior force.

II. Law of Acceleration: Change in a body's motion is proportional to the magnitude of any force acting upon it and in the exact direction of the applied force.

III. Law of Reaction: Every acting force is always opposed by an equal and opposite reacting force.

Model rockets—and everything else in the universe that moves—obey all three of these Laws of Motion.

At first glance it does not seem rational that if you push a body, it will keep on going forever and ever at the same speed and in the same direction that you pushed it. Our everyday experience tells us that the body slows down and stops. But the reason it slows down and stops is the application of an external force, usually friction, acting upon the body to change its velocity in accordance with the Law of Acceleration.

A model rocket is one of the few easily accessible objects that obeys Newton's three Laws of Motion in such a straightforward way.

A model rocket's flight is primarily affected by the Second Law, which can be stated mathematically as follows:

$$F = ma \qquad\qquad (1)$$

where F = the applied force (from the rocket motor or from gravity), m = the mass of the model rocket, and a = the resulting acceleration.

If the applied force is doubled (that is, if the thrust is doubled) and the mass is kept constant (that is, the weight of the model remains unchanged), the model's acceleration will be doubled. If the force remains the same and the mass (weight) is doubled, the acceleration will be reduced by one-half.

From the Laws of Motion and the basic equations of velocity and acceleration, I've derived the following general equations relating to motion, distance traveled, velocity, acceleration, and time. If you're interested in how they were derived, you'll find the derivations in any high school physics text. Otherwise, just use them as they are here to help you determine the flight characteristics of a model rocket. Actually, they are greatly simplified, since few of you probably know calculus yet.

$$s = vt \qquad\qquad (2)$$

$$V_{av} = \frac{v_1 + v_2}{2} \qquad\qquad (3)$$

$$a = \frac{v_2 - v_1}{t} \qquad (4)$$

$$s = v_1 t + \frac{at^2}{2} \qquad (5)$$

$$2as = v_2^2 - v_1^2 \qquad (6)$$

where s = distance, v = velocity, v_1 = velocity at start of time period, v_2 = velocity at end of time period, V_{av} = average velocity during time period, a = acceleration, and t = length of time period.

Appendix II at the back of this book shows how you can use these equations to compute acceleration, velocity, and distance traveled by a model rocket.

But a model rocketeer is primarily interested in only a few things, such as altitude and coasting time. These flight characteristics can be computed for the drag-free condition by a very simple method if you have three pieces of information about your model rocket:
1. You must know the *total impulse* of the model rocket motor.
2. You must know the *burnout weight* of the model—that is, its takeoff weight minus the propellant weight.
3. You must know the *duration* of the model rocket motor thrust.

You can get the total impulse from the type code of the model rocket motor or from the manufacturer's specifications.

The weight of your model rocket can be determined by weighing it on a laboratory balance at school or on a postal scale. Weigh it with a loaded motor and wadding installed, in the exact condition it will be at lift-off.

The duration of the model rocket motor thrust is obtained from the motor manufacturer's specifications that are given on a sheet of paper packed with the motor.

Let's run through an example here. Suppose that your model rocket has the following characteristics:

Total impulse of motor (I_t) = 2.25 newton-seconds
= 0.506 pound-seconds
Duration of motor (t_b) = 0.32 seconds
Lift-off weight of model (W_0) = 30.1 grams
= 1.06 ounces
= 0.066 pounds

Propellant weight (W_p) = 3.12 grams
= 0.11 ounces
= 0.0069 pounds
Burnout weight of model (W_1) = $W_0 - W_p$
= 30.1 − 3.12 = 26.98 grams
= 0.95 ounces
= 0.0594 pounds

Note that I have started with all units in the metric system and have converted into the English system. All units are in terms of *pounds* and *seconds* in the English system. If I had calculated in the metric system, I would have converted to *kilograms* and *seconds*.

Recall that by definition the total impulse is the total change in the momentum of a body. Momentum is mass times velocity. Written in mathematical equation form, this is:

$$I_t = m_1 v_1 - m_0 v_0 \tag{7}$$

where I_t = total impulse, m_1 = mass at burnout, v_1 = velocity at burnout, m_o = mass at lift-off, and v_o = velocity at lift-off.

Since at zero time, or the instant of launching, the model rocket's velocity is zero—it's sitting on the launch pad—the equation reduces to:

$$I_t = m_1 v_1 \tag{8}$$

Now, when you weighed your model, you actually recorded the *force* with which it was being pulled by gravity toward the earth's center. To determine the model's *mass*, you must take its weight (which is not its mass) and divide it by the acceleration of gravity (32.2 feet per second per second):

$$m = \frac{W}{g} \tag{9}$$

So Equation (8) becomes:

$$I_t = \frac{W_1 v_1}{g} \tag{10}$$

Transposing to get v_1 over to the left side by itself, we get:

$$v_1 = \frac{I_t\, g}{W_1} \tag{11}$$

The term v_1 is equal to the velocity at the end of the impulse; therefore, it is the burnout velocity of the model rocket. It is also the maximum velocity the model will attain during its powered flight and its coasting flight. Therefore, we call it V_{max}. And for our hypothetical model rocket in the example we can calculate:

$$V_{max} = \frac{0.506 \times 32.2}{0.0594}$$

$$= \frac{16.29}{0.0594}$$

$$= 274.3 \text{ feet per second}$$

That is the maximum velocity attained by the model rocket. It is equal to 187 miles per hour.

Remember that I said model rockets were the world's fastest models? And this example was propelled by a reasonably low-powered Type A motor!

We must now compute how high the model is at burnout (S_p). We can use Equation (3) and Equation (2) given earlier.

$$V_{av} = \frac{v_1 + v_2}{2}$$

$$= \frac{0 + 274.3}{2}$$

$$= 137.15 \text{ feet per second}$$

$$S_p = vt$$
$$= V_{av}\, t_b$$
$$= 137.15 \times 0.32$$
$$= 43.88 \text{ feet}$$

Now you know why recovery devices are not deployed at burnout of the model rocket motor! At burnout our hypothetical model is only about 44 feet in the air and is traveling at 187 miles per hour! It hasn't neared its maximum altitude, and it is moving at such speed that it would tear any recovery device to pieces.

When the motor thrust drops to zero at burnout and the time delay begins to work, the model enters the coasting phase of flight. It coasts upward to maximum, or peak, altitude (apogee), trading its momentum (velocity) for altitude. During coasting flight only gravity and aerodynamic drag forces are acting on the model. Since we are ignoring aerodynamic drag for this example, only gravity is acting. The model is actually "falling upward" in a gravity field; it is in zero-g, or weightlessness. It is being acted upon by the acceleration of earth's gravity, 32.2 feet/sec². We can now go to Equation (6) and find out how far it will coast:

$$2as = v_2^2 - v_1^2 \qquad\qquad (6)$$

Since $v_1 = 0$, $2as = v_2^2$. Therefore:

$$s = \frac{v_2^2}{2a}$$

where s = altitude, v_2 = maximum velocity, and a = acceleration. This results in the coasting altitude (S_c) calculation:

$$S_c = \frac{V_{max}^2}{2g}$$

$$= \frac{274.3^2}{64.4}$$

$$= \frac{75240.49}{64.4}$$

$$= 1,168.3 \text{ feet}$$

The maximum altitude (S_t) will then be the burnout altitude (S_p) plus the coasting altitude (S_c):

$$S_t = S_p + S_c$$
$$= 43.88 + 1,168.3$$
$$= 1,212.2 \text{ feet}$$

But what would happen if we put a Type B motor with twice the total impulse into the model and flew it again? The new model parameters would be:

131

> **Total impulse = 4.5 newton-seconds**
> **= 1.01 pound-seconds**
> **Duration of motor = 0.83 seconds**
> **Lift-off weight = 36 grams**
> **= 1.27 ounces**
> **= 0.079 pounds**
> **Propellant weight = 6.24 grams**
> **= 0.22 ounces**
> **= 0.0137 pounds**
> **Burnout weight = 29.76 grams**
> **= 1.05 ounces**
> **= 0.0656 pounds**

Running through the calculations quickly, we find:

$$V_{max} = \frac{1.01 \times 32.2}{0.0656}$$
$$= 495.76 \text{ feet per second}$$

$$V_{av} = \frac{495.76}{2}$$
$$= 247.88 \text{ feet per second}$$

$$S_p = 247.88 \times 0.83$$
$$= 205.74 \text{ feet}$$

$$S_c = \frac{495.76^2}{64.4}$$

$$= \frac{245777.97}{64.4}$$
$$= 3,816.43 \text{ feet}$$

$$S_t = 205.74 + 3,816.43$$
$$= 4,022.17 \text{ feet}$$

Quite a difference! We *doubled* the model's total impulse, but its performance didn't simply double, did it? The model was slightly heavier at lift-off because of the additional propellant of the Type B motor, but the added weight was more than offset by the motor's longer burning time. This pushed up the burnout altitude and also caused a greater burnout velocity. When we compare the two performances, the following points stand out:

1. If we *double* the total impulse, the maximum velocity increases almost *four times*. In other words, the maximum velocity increases as the *square* of the total impulse increases.

2. If we *double* the total impulse, the burnout altitude increases almost *four times*. Again, the burnout altitude increases as the *square* of the increase in total impulse.

3. If we *double* the total impulse, the maximum altitude increases about *four times*, too.

In summary, the performance increases as a function of the *square* of the increase in total impulse! This is an important relationship to remember.

By running off another set of calculations, you can also see for yourself that any increase in the weight of the model will decrease the altitude performance.

Although they work out mathematically, the numbers we have just calculated don't seem to jibe with reality, do they? From our own experience flying small models of this sort with Type A and Type B motors, we know that they do not go to altitudes of over 1,000 and 4,000 feet. Generally, they go to about 500 and 1,000 feet.

What's wrong? The fact that we deliberately ignored the effects of aerodynamic drag in our flight calculations. And we can now see very dramatically that aerodynamic drag is a major force in model rocketry. It is also obvious that we *must* take aerodynamic drag into account when calculating flight performance.

Believe it or not, air is considered to be a fluid. And the amount of air drag experienced by a model rocket can be calculated by an equation from the science of fluid dynamics.

The basic drag equation is:

$$D = 0.5\rho V^2 C_d A \tag{12}$$

where D = drag force, ρ = air density, V = velocity of the model through the air, or the air past the model, C_d = a dimensionless number called the drag coefficient, and A = the frontal area of the model.

The drag equation tells us:

1. Air drag increases as the air density increases. How can you change the density of the air to reduce drag? By going to a higher altitude where there is less air, or by flying on a hot day because air is less dense when it is hot.

2. Air drag increases as the *square* of the velocity. Double the velocity, and the drag force goes up *four times*. The faster the model goes, the greater the drag force becomes. See Figure 7-4.

3. Air drag increases directly as the drag coefficient increases. We will discuss this point in detail because it is something you can work with.

4. Air drag increases directly as the frontal area increases.

Model rocketeers can't do too much about the air density—except try for maximum altitude by flying from the top of Pike's Peak instead of at sea level, or by flying in the desert at 110° F. instead of in Minnesota at -20° F. The amount that drag force is changed by changes in launch altitude and air temperature is shown in Tables 7-A and 7-B.

The value of the drag coefficient C_d depends upon many factors. Primarily, it is a function of the shape of the model.

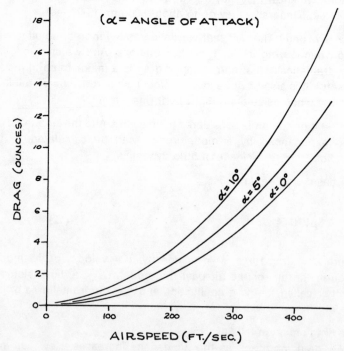

Figure 7-4: Drag versus airspeed for a model rocket at varying angles of attack.

Table 7

A. Altitude Correction Table

For computing change in air density as a function of elevation of launcher above sea level.

Elevation (in feet)	Multiply by
0	1.0000
1,000	.9710
2,000	.9428
3,000	.9151
4,000	.8881
5,000	.8616
6,000	.8358
7,000	.8106
8,000	.7859
9,000	.7619
10,000	.7384

B. Temperature Correction Table

For computing change in air density as a function of air temperature at launcher.

Temperature (in °F.)	Multiply by
30	1.0590
35	1.0486
40	1.0380
45	1.0277
50	1.0177
55	1.0078
60	.9980
65	.9885
70	.9792
75	.9700
80	.9610
85	.9522
90	.9435
95	.9350
100	.9266

This includes the shape of the nose; whether or not there are any transitions or changes in body diameter; the number, planform, airfoil, and tip shape of the fins; the location and size of the launch lug, if any; and the smoothness of the surface finish of the model. (The size of the model is taken into account in the frontal area term of the equation.)

The drag coefficient is not constant. It also depends upon the model's angle of attack, the angle between the long axis of the model and the direction of the air flowing past. See Figure 7-5. For most model rocket shapes the drag coefficient increases with increasing angle of attack, as shown in Figure 7-6. As you can see, the frontal area of the model presented to the oncoming airflow also increases with angle of attack, and this increases the value of A in the drag equation.

What do these facts mean? If a model wobbles in flight, thereby flying at different angles of attack, its drag force will be greater than that of a model that slips through the air smoothly with little or no wobble. This is only one reason why it is important to design and build a stable model rocket.

The methods used to alter the drag coefficient will be discussed in detail in the chapter on aerodynamic stability and shape.

Now, how do we work the aerodynamic drag force into the flight equations of a model rocket? Answer: With great difficulty. Before the digital computer came upon the scene to help, this calculation was long, tedious, and painstaking. This is because the equations get hairy, to use engineering slang.

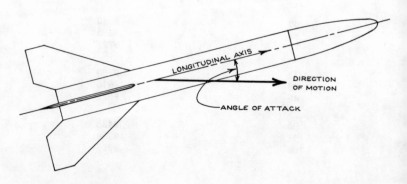

Figure 7-5: Angle of attack.

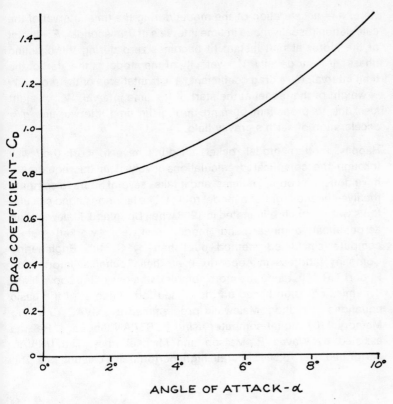

Figure 7-6: Drag coefficient (C_d) of a typical model rocket as a function of angle of attack.

Look at the drag equation, and you will immediately see that the drag force not only opposes the motion of the model through the air, but also changes as the model's velocity changes. When the model takes off, the velocity is low, the drag force is low, and the model behaves as though there were no drag. Then, as the velocity begins to build up, the drag force increases. This retards the model more. It is a very complex interrelationship. If you know calculus, the mathematics are simple. But the objective of this handbook is to make things easy for people who are not necessarily math sharks or computer engineers.

Assuming a vertical flight, the acceleration of the model at any instant during its powered or coasting flight can be determined from the equation:

$$a = \frac{F - (0.5\rho V^2\, C_d A)}{W - W_0}\, g \qquad (13)$$

137

Here, a = acceleration of the model during the time interval of the calculation (usually made in time intervals of 0.1 seconds), F = thrust of the motor at that instant (it becomes zero during the coasting phase), ρ = air density, V = velocity of the model at the start of the time interval, C_d = drag coefficient, A = frontal area of the model, W = weight of the model at the start of the time interval, W_o = weight loss due to propellant burning during the time interval, and g = acceleration of earth's gravity field.

Happily, modern model rocketeers don't have to work their way through the complicated calculations involved in this equation. I have done it a couple of times, and it takes several hours to compute the five-second flight of a model rocket. The laborious hand calculations were virtually eliminated in 1968 when Douglas J. Malewicki, an aeronautical engineer and model rocketeer, worked out a computer-calculated method published by Centuri Engineering Company (address in Appendix I) as their Technical Information Report TIR-100. Using the more complex versions of the above flight dynamics equation based on the closed-form integral of the basic equations of motion, Malewicki programmed a UNIVAC Thin-Film Memory 1107 digital computer using FORTRAN language. He was assisted by Wayne R. Matson and Michael Poss. The UNIVAC performed over 13,500 separate flight performance calculations. The

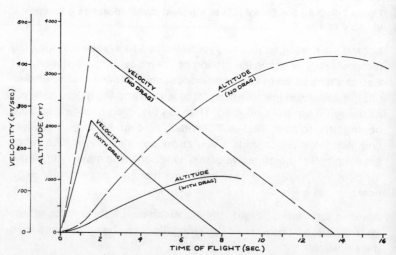

Figure 7-7: Comparison of computed drag-free flight performance with realistic performance taking air drag into account.

results yielded a set of charts for each model rocket motor type. All a model rocketeer needs to know is the motor type, the model lift-off weight, the body diameter, and the drag coefficient. With this information the rocketeer can consult two charts and predict with less than 1% error the peak altitude of any model rocket.

This outstanding contribution to model rocketry has permitted model rocketeers to engage in a new type of contest known as predicted altitude. The objective of this contest is to predict before the flight the altitude to which one's model will fly. The winner is the person whose prediction comes closest to the actual tracked altitude of his model. At national model rocket meets the winners usually make perfect predictions for flights of 500 to 1,000 feet.

If we compare the flights of model rockets drag-free and with aerodynamic drag taken into account, the comparison comes out as shown in Figure 7-7. If we make a careful analysis of such a comparison, we can draw the following conclusions about the effects of aerodynamic drag on model rockets:

1. The drag of a well-streamlined model rocket (one with a low value of C_d) can lower the burnout altitude to 40% to 60% of the computed drag-free value.

2. Drag lowers the computed drag-free altitude of a model rocket by more than 50% in nearly all cases; it can actually reduce drag-free altitude by as much as 95% in the case of a poorly built, unstreamlined garbage model.

3. Drag forces on a model rocket become very large at velocities of 150 feet per second or more, requiring very rugged construction for high-performance models to be propelled by NAR Type D, Type E, and Type F motors.

4. The highest drag force occurs within a second following burnout and rapidly becomes less as the model coasts up to peak altitude. After burnout-plus-one-second the drag force drops to 60% to 80% of its burnout value.

There is another interesting fact hidden in these equations. The drag-free calculations would lead you to believe that the lighter the model, the higher it will go. This is not so when air drag is taken into account. You cannot eliminate the factors relating to drag coefficient and frontal area. The basic equation (Equation 13) shows that C_d is divided by weight. Therefore, you eventually reach a point where your ultralight model rocket is just like a feather—and you can't throw a feather very far! The model's area-to-weight ratio gets to be so great that finally the aerodynamic drag force completely overwhelms the momentum (mass times velocity).

On the other hand, if you make a model rocket too heavy, the thrust from the motor will not be able to accelerate it to a high velocity, and the peak altitude will be reduced.

So a model rocket designer gives a little here, takes a little there, and tries to reach some sort of a happy compromise. For some reason the early model rocket designs were right at or very near the optimum trade-off between weight and drag. This was a very fortunate circumstance because it permitted the historic models to operate satisfactorily, giving model rocketry time to develop the nuts-and-bolts of its technology before having to get into the complexities of weight-area trade-offs.

Analysis of flight dynamics and trajectories is a favorite subject for college undergraduates who have been model rocketeers and have carried their hobby into college with them. The Massachusetts Institute of Technology's Model Rocket Society has done a tremendous amount of work in this area using digital computers. Cadets at the Air Force Academy have utilized their wind tunnels for further studies into the subject.

Yet this is an area that junior high school students can work in with minimal mathematical tools and abilities. You can predict with great accuracy the altitudes, accelerations, and velocities your models will attain. These numbers are of great help in designing your model rockets properly. Besides, it is very exciting to see the model rocket you have designed and "flown on paper" perform in the real world just as you calculated. It is a dynamic example of how engineers and scientists can predict the future by knowing how the universe works.

Stability and Shapes

In the preceding chapter we have seen that the earth's atmosphere, the air, plays a major role in the flight of a model rocket. Aerodynamic drag reduces performance. But, by utilizing proper model design, we can *use* the air to stabilize the model, to keep it going in the intended direction, and to make sure that it flies predictably. We can also decrease the aerodynamic drag and increase the performance of a model rocket by understanding how the air flows around it and creates the drag force.

We must always keep in mind that a model rocket is a free body in space after it leaves the launch rod. It is not attached to the ground, and the forces acting upon it in flight cannot be easily duplicated on the ground.

In discussing the performance of a model rocket in flight, we acted as observers on the ground, watching the model fly with reference to the ground. We were standing still and watching the model move. To better understand aerodynamic stability and aerodynamic shapes, we must change our point of view and travel with the model in its flight through the air. In other words, we must become theoretical passengers in order to see and understand better what is happening.

A model rocket in flight can move in *eight* different ways. This is technically known as eight degrees of freedom. For simplicity and ease in considering the motions, we can reduce any motion of a model rocket in flight to a combination of one or more of the eight basic motions shown in Figure 8-2.

Figure 8-1: The size and shape of model rockets determine most of the air drag that they will encounter in flight. Model rocketeers have developed many different shapes for various purposes.

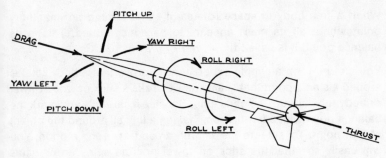

Figure 8-2: The eight degrees of freedom of a model rocket in flight. Only two—thrust and drag—are linear; the other six are rotational.

Thrust moves the rocket forward. It comes from the model rocket motor.

Drag opposes the thrust force and attempts to slow the model.

(The *gravity* force can come from any direction, depending upon the attitude and flight direction of the model; we have already taken that force into account in calculating the overall flight dynamics of the last chapter.)

Yaw is a swinging motion of the nose to left or right.

Pitch, an up or down motion of the nose, is similar to the yaw motion. Because a simple model rocket is the same in both the pitch and yaw aspects, the two motions are usually lumped together as a *pitch* motion. But this is not true for most boost-gliders and rocket gliders, as we shall see in a later chapter.

Roll is a rotational motion whereby the model spins right or left about its long axis.

Thrust, drag, and gravity forces are linear; they produce linear motions called translations.

Pitch, yaw, and roll are rotational motions. Anything that rotates must have an axis of rotation, an imaginary line around which it revolves. The earth, for example, has a rotational axis running through the North and South Poles. A model rocket has a roll axis that is an imaginary line running through the tip of the nose down along the center of the model and out the nozzle of the motor. There is also a pitch axis and a yaw axis, imaginary lines through some points in the model. Where are these points? How do we find them?

When a free body in space rotates, it spins around an imaginary point where all its mass appears to be concentrated. This is its balance point. It is called the *center of gravity*, or *CG*.

You can perform a simple experiment to prove that a body rotates around a single point that is also its *CG*. Take a stick or wood dowel or body tube about 18 inches long. Balance it carefully and mark the balance point with a felt-tip pen, making a line all around the object at that point. Toss it into the air with an end-over-end motion. You will easily see that the stick or dowel or tube spins around this balance point as an axis. No matter how you throw the stick to get an end-over-end motion, it will always rotate around this point.

Now put some putty or plasticene clay firmly on one end of the stick. Rebalance the stick and mark the new *CG* point, using a different-colored felt-tip pen. The *CG* will be in a new location, closer to the end of the stick on which you put the additional weight. Again throw the stick end-over-end, and you will see that it rotates around the new balance point.

A model rocket in flight tends to rotate around its *CG* in the pitch, yaw, and roll axes. Why be concerned about these motions? Because if a model rotates around the pitch or yaw axes without interference, it's going to change its direction of flight! And you want that bird to go right where you pointed the launch rod—straight up. If it leaves the launch rod only to spin around its pitch and yaw axes, its angle of attack (and therefore its aerodynamic drag) will increase and the model will go in a direction other than the way the launch

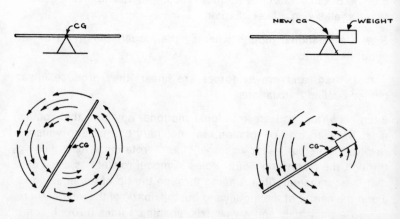

Figure 8–3: Rotation of a body around its center of gravity.

rod is pointed. These other directions can often be very erratic and unpredictable, to say the least.

What can cause the model to rotate around its pitch-yaw axis? As Sir Isaac Newton said, in a voice that calls down over the centuries to the spacemen of today: A change of motion in a body can be produced only by an external force acting on that body.

There are many external forces that can cause rotational motions in a model rocket—gusts of wind blowing at the instant of launch, winds blowing at various altitudes during flight, fins crookedly positioned on the body tube, off-center thrust from the motor, irregularities in model construction, and many others. No matter how perfectly you build a model and no matter how carefully you try to launch it under ideal conditions, there will always be some tiny forces that will begin to produce pitch-yaw rotational forces the instant the model leaves the launch rod. You cannot eliminate them.

Therefore, your model must incorporate some sort of stabilization device that will overcome these forces, damp them out, and restore the model to its intended flight direction. Furthermore, this must be done very quickly—in a fraction of a second. Instantly a force must be created to oppose the rotational force.

Most big space rockets at the Cape counteract rotational forces with an automatic control system, an autopilot that uses gyros to sense the rotation in the pitch-yaw axes, and then sends electrical signals to a computer and tilts the rocket motors by means of hydraulic pistons to correct the disturbing force. These control systems are very heavy, very large, very complicated, and very expensive. Although some model rocketeers have built primitive gyros for their birds, most model rockets get their stabilizing and restoring forces simply from the air rushing past the model and acting upon aerodynamic surfaces—the fins.

When a moving stream of air strikes a surface broadside (at an angle of attack of 90 degrees) or even at a slight angle, it produces a high pressure on one side of the surface and a low pressure on the other side of the surface, as shown in Figure 8-4. This pressure difference produces a *drag* force opposite to the motion of the airstream and a *lift* force that is at right angles (90 degrees) to the surface. The higher the angle of attack (see the preceding chapter if you made the mistake of glossing over this subject), the greater the lift and drag forces—up to the point where the surface stalls. At the stall point the air breaks away from the low-pressure surface, the lift force drops drastically, and the drag force rises tremendously.

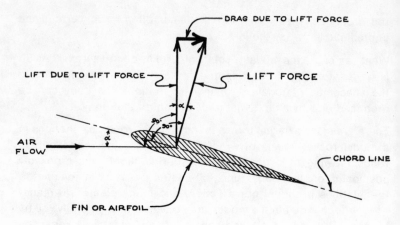

Figure 8-4: Lift force and drag force on a fin (shown in cross section) at an angle of attack.

By properly positioning the fins on the model rocket, by making them the right size, and by giving them the right shape, we can use this lift-drag force as a stabilizing force to offset pitch-yaw rotational disturbances.

As we saw earlier with our tumbling stick, we can make the simplifying assumption that all the weight of a body is concentrated at its *CG*. It certainly acts that way. We can now extend this concept to *any force* acting upon the body, including the aerodynamic lift-drag force we have just introduced. Therefore, we can think of the body as having a point at which all air pressure forces act. We call this point the *center of pressure*, or simply the *CP*.

The function of *CP* can be illustrated by another experiment with the wooden stick. If we grasp the stick at its balance point between a

Figure 8-5: The stick-and-vane model under various conditions described in the text.

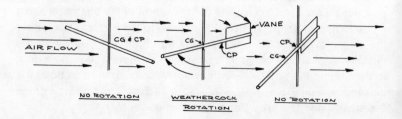

pair of pointed pivots, as shown in Figure 8-5, and hold it in the moving air from an air-conditioning duct (not from a fan because of the turbulence) or out of the window of a car traveling at about 20 miles per hour, we will see that the stick does not rotate in the pivots due to the air pressure. No matter how you hold the stick in the airstream, it does not rotate. Obviously, there are no off-center forces to make it rotate, so the *CG* and the *CP* must be the same.

Now glue or staple a piece of stiff cardboard to one end of the stick. Because the weight of the cardboard vane will change the *CG* point, balance the stick again. Pick up the stick with the pivots at the new *CG*, and place it in the moving air again. The stick will immediately swing around with the vane downwind and the stick pointed directly into the wind. Congratulations! You've just reinvented the weather-vane!

The presence of the cardboard vane produced more air pressure force on its end of the stick. This air pressure force was caused by the lift-drag of the vane.

If you push the stick slightly to one side with your finger, you will discover that the air pressure force becomes greater as you displace the stick from the nose-into-wind condition. There is very little force when the angle of attack is low, but more force as the angle of attack increases. This is the restoring force that will stabilize our model rockets.

If we now start to move the pivot point closer to the vane end of the stick, we will eventually find the point where the stick will no longer pivot into the wind. This is where the air pressure forces on both ends of the stick are equal. We have just located the center of pressure (*CP*) of our stick-vane model.

We can replace the stick and vane with a model rocket, and it will behave in the same manner. When picked up and supported by pivots at its *CG*, it will always point into the airstream. This tells us that its *CP* is *behind* its *CG*, making it a stable model.

What would happen if we picked up the model or the stick at a point behind the *CP*? Very interesting! It would try to fly tail-first and would gyrate pretty wildly. A real model rocket under thrust does not want to fly tail-first—in no way by any manner or means! It is an unstable model.

As shown in Figure 8-6, there are basically three stability conditions for a model rocket—and only one of them is desirable. In summary they are:

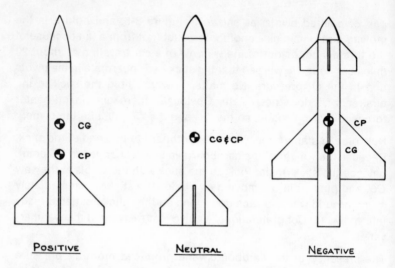

POSITIVE   NEUTRAL   NEGATIVE

Figure 8–6: The three stability conditions with their *CG-CP* relationships.

1. *Positive stability*, where the *CG* is ahead of the *CP*. The model is nose-heavy. It has large fins set well back on the body tube. It will fly straight when launched and will weathercock into the wind at launch.

2. *Neutral stability*, where the *CG* and the *CP* lie at the same location on the model. This might be caused by a lightweight nose or by fins that are too small, or both. There are no stabilizing and restoring forces present in the model during flight. It is free to wander anywhere in the sky, and some of its wanderings may be wild and unpredictable. It may also become stable or completely unstable at any moment—while pointed in any direction!

3. *Negative stability*, where the *CG* lies behind the *CP*. In this case the aerodynamic forces of the fins try to make the model fly tail-first, which it does not want to do under power. Once the nose swings in pitch or yaw after leaving the launch rod, a force keeps it swinging. The model usually pinwheels end-over-end and winds up going nowhere except to flop on the ground.

Remember the positive stability condition by the mnemonic: *G* before *P*. This is the alphabetical rule, because the *CG* comes before the *CP* in a flyable model.

You should *never* fly a model rocket until you have determined its *CG-CP* relationship to ensure that it will be stable in flight. The best way to make this determination is by the swing test. Tie a 4-foot length of string around the body tube of the model so that the model with a loaded motor installed hangs horizontally from the string. In other words, wrap the string around the model at its *CG*. Hold the string in place with a bit of tape.

Then, making sure that nobody is in the way, start to swing the model around your head in a horizontal circle. The longer the string, the more valid this test. If the model is stable, its nose will point in the direction of motion as you swing it. If the nose points elsewhere, you must add some nose weight to bring the *CG* forward.

It is interesting—and perhaps dismaying—to learn that the *CP* of a model rocket depends on its angle of attack! The model may be stable if it swings to a small angle of attack. But if it swings to a high angle of attack, or if a strong gust of wind hits it in flight, it may not be able to recover its stability. This is because the *CP* of a model rocket moves forward with increasing angle of attack. It reaches its most forward point when the angle of attack is 90 degrees. Some model rocket designs have less *CP* movement than others.

How does a model rocket designer handle this mess? Quite easily.

In the early days of model rocketry we computed the *CP* location by letting it equal the center of lateral area. To determine this point, we

Figure 8–7: Model rocketeer making the swing test of a model to determine stability.

Figure 8–8: Determining the *CP* of a model using the cardboard cutout method.

made a cardboard cutout of the side-view shape of the model. When we balanced this cardboard cutout, it gave us the center of lateral area, which we used as the *CP*. Actually, the center of lateral area is indeed the *CP* of the model *if* it is flying at an angle of attack of 90 degrees, the worst possible condition.

All of the models we designed using this method flew. But there were a few models that just "happened" and were not designed. They flew well even though the cardboard cutout method of *CP* location indicated they should be unstable. We knew from NACA and NASA reports that aerodynamicists had a method of computing the *CP*, but it was a very complicated method that nobody seemed able to apply to model rockets.

The breakthrough came in 1966 when James S. Barrowman, a former president of the National Association of Rocketry and a professional aerodynamic engineer with the Sounding Rocket Branch of NASA's Goddard Space Flight Center, presented a simplified technique for computing the actual *CP* of a model rocket at low angles of attack.

The Barrowman Method has now been refined so that you don't need any math more advanced than simple arithmetic to calculate the *CP* of the most complex model rocket you can think of. One method that I have developed for my own design work is presented in the Appendix. To use it, you must have Technical Information Report TIR-33 from Centuri Engineering Company. It is a slick and quick way to do what used to be an impossible job. If you are a serious model rocket designer who wants to shave the stability margin as close as possible, obtain Centuri TIR-33 and cherish it.

Otherwise, you can get by just fine by using the old cardboard cutout method. Your models will have too much fin area and will have a tendency to weathercock excessively in the wind, but they will fly stably.

How much stability should you have in a model rocket? How far behind the *CG* should the *CP* be?

It is generally agreed among advanced model rocketeers that the *CP* should be no less than one body diameter behind the *CG*. In other words, if the body tube diameter is 1.04 inches, the *CP* should be at least 1.04 inches behind the *CG*. This is known as *one-caliber stability*. In gunnery terms "caliber" refers to body diameter, and the word comes to us from the days when rockets were part of the artillery of armies.

Technically, one-caliber stability is all you really need for most sporting models. It allows for the forward movement of the *CP* with increasing angle of attack up to a reasonable limit beyond which the model probably won't go anyway. If you have more than about two-caliber stability, your model may be overstable and suffer from excessive weathercocking (which may be something that you want in a parachute duration type of contest model).

What do you do if your model rocket checks out as neutrally stable or negatively stable? That depends on the model. It may be possible to increase the fin area or to move the fins back; this moves the *CP* rearward. Such an approach may not be possible with a scale model, for example; then weight must be added to the nose to bring the *CG* forward. Sometimes you must do a little of both to obtain optimum weight and the least amount of fin area. This is one of the things that makes a model rocket designer's work so fascinating. There are so many trade-offs that can be made.

For many years—and even now in some parts of the world where model rocketry is new—model rocketeers believed that they should make the smallest, lightest model rocket possible in order to achieve the highest possible altitude. As we have seen in a previous chapter, there is at least one serious flaw in this reasoning because it is possible to make a model rocket so lightweight that it tries to fly like a feather. This same sort of make-it-small design philosophy resulted in some very short, very squat, very fat little model rockets. Essentially, they were barely large enough to enclose the motor, a very small recovery streamer, and a nose cone. To obtain the proper stability, they had fins that were sharply swept back behind the

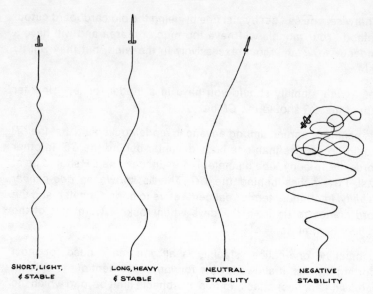

SHORT, LIGHT, & STABLE   LONG, HEAVY & STABLE   NEUTRAL STABILITY   NEGATIVE STABILITY

Figure 8–9: Flight paths of model rockets with varying static and dynamic stability characteristics.

model. True, they had one-caliber stability. But all of the major weights of the model—motor, nose, recovery device—were located very close together.

In flight these models wobbled back and forth very rapidly as they ascended. We discussed earlier that the drag increases as the angle of attack increases. Therefore, these squatty models experienced a great deal of drag because of their constant and rapid changes of angle of attack.

On the other hand, many of us had started flying long, slender models that perhaps weighed a bit more, but would slither into the air with very little wobble.

The basic performance is shown diagrammatically in Figure 8-9.

To understand why the two basic designs function as they do, let's go back to our stick model held between pivots out of a car window. Make a short stick about 1 inch in diameter and about 4 inches long. Put a cardboard vane 2 inches by 4 inches on the stick, as shown in Figure 8-10. Make a longer stick about 1 inch in diameter and 12 inches long. Cut a cardboard vane 1 inch by 2 inches and staple it to the stick, as shown.

Grasp the short stick by the pivots at the *CG* and put it in the airstream. Displace it in the yaw direction with your finger. It is a stable model and will recover. But watch how it does it. The air hits the vane and produces a restoring force due to the lift-drag force on the vane. The stick starts to swing back to zero angle of attack. It does so very quickly because all of its mass is concentrated near its *CG*. Very little inertia is involved. The stick will reach its maximum rotational velocity when it hits zero angle of attack, and it will not stop swinging. It will swing through zero angle of attack and take up an angle of attack on the other side of zero. The stabilizing situation is reversed, and the model will start to swing back. It will oscillate back and forth on either side of the zero angle of attack point. The oscillations will be very rapid, and several will be required before they damp out, or stop swinging.

Now hold the long stick model in the airstream in the same way. It is also stable. Displace it in the yaw direction as you did the short model. When released, it will also start swinging back, but it will take longer to do so. Its rotational velocity as it passes through zero angle of attack will be much less. It may also swing to the other side. But its swings will be slow and will damp out after perhaps only one or two passes through zero angle of attack.

These actions are caused by very complex phenomena grouped together under the general classification of dynamic stability. Just because we can hang a name on it doesn't mean that we understand it any better. Basically, the difference between static stability and dynamic stability is that of balancing a nonmoving device versus balancing a moving machine.

Figure 8–10: Visualization of dynamic stability with the stick-and-vane model.

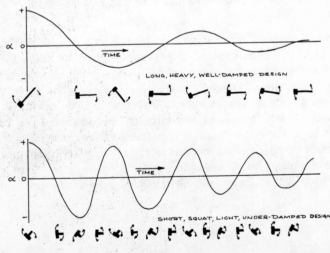

It is perfectly possible for a model rocket to have its *CG* and *CP* in the proper location with respect to each other, thereby possessing static stability, while actually being unstable in flight. There are several types of dynamic stability problems ordinarily encountered by model rocketeers:

1. A statically stable but dynamically underdamped model. This is the short, squat, fat little model that wobbles excessively as it flies.

2. A statically stable but dynamically unstable model. In this condition the *CP* and *CG* may be located properly, but the model is short and heavy and has fins that are too small to produce enough restoring force to return the model quickly to zero angle of attack. The fins are large enough to bring the *CP* behind the *CG*, but not large enough to exert sufficient force to make the model react quickly enough. By the time the fins cause sufficient reaction or restoring force, the model is doing something else.

3. A statically stable but dynamically overdamped model. In this condition you might have a long, skinny model with fins that are far too large. When the model weathercocks or rotates in the pitch-yaw axes, the fins produce too much force and stabilize the model too quickly, causing it to fly as though it were almost neutrally stable.

4. A model with pitch-roll coupling. This is a weird form of instability that can really frustrate you if you don't know about it. In this condition the model has roll, caused by fin misalignments, nose misalignment, or even motor misalignment. The frequency of the roll, at some point in the flight, will closely match the frequency of oscillation in the pitch-yaw axes. The model may start by exhibiting a coning motion in which it is spinning at the same time it is rotating in a conical motion about the pitch-yaw axes. Finally, the roll frequency couples with the pitch-yaw wobble frequency, and the model becomes completely unstable. This problem often occurs in full-scale sounding rockets, and I have had it happen to me a couple of times with new designs.

I first mentioned these dynamic stability problems in an earlier edition of this book, explaining that they were very complex and not at all well-understood even by professional rocket engineers. As a result, Gordon K. Mandell tackled the subject while he was an undergraduate at Massachusetts Institute of Technology. He improved the MIT low-speed, low-turbulence wind tunnel and developed some ingenious measuring instruments. He made some mathematical simplifying assumptions relating to the very complex equations of motion, and finally succeeded in linearizing the equations to the point where they could be useful to advanced model rocket designers. Mandell didn't rest his case on mathematical theory alone; he checked his results and verified his assumptions by actual tests of model rockets under carefully controlled conditions in the MIT wind tunnels. His thesis was first published in the old *Model Rocketry* magazine in 1968. The work is now available in its entirety in the very sophisticated book *Topics in Advanced Model Rocketry* by Mandell, Caporaso, and Bengen, published by the MIT Press in 1973.

Basically, Mandell discovered that most of our model rocket designs were overdesigned on the safe side. This was not surprising, since most of us tend to be a bit conservative in design because of all the unknown factors at work. No real rules of thumb have yet been worked out from Mandell's classic experiments, but perhaps we may be able to report on some in a future edition. Briefly, some points of design brought forth by Mandell are as follows:

1. Maintain a length-to-diameter ratio of no less than 10 to 1 to provide adequate damping.

2. Maintain a static stability margin between one caliber and two calibers to prevent overdamping, but don't go less than one caliber.

3. Hold the roll rate as low as possible to prevent pitch-roll coupling, and try to get the model to fly with no roll by careful alignment of the fins.

4. If you have to increase the linear dimensions of the fins to get proper static stability, increase the *span* dimension (the dimension outward at right angles to the body) because this will increase the restoring force rather than increase the distance of the *CP* from the nose tip.

Static stability has been pretty well whipped, and Mandell has made significant progress toward solutions of the problems in dynamic stability. But we still need rational rules of thumb backed up by solid theoretical and experimental work, so that the average model rocketeer can apply Mandell's findings to his own designs using only simple arithmetic and graphs. The problem could be tackled by a high school senior, but more easily by a college undergraduate. The person who resolves it successfully will be enshrined next to Barrowman and Mandell when rocketeers talk about model rocket stability.

Don't think this work is kid stuff! Professional rocket designers have started to use Mandell's findings because they are applicable to full-scale sounding rockets, too! It took a model rocketeer to get things simplified to a workable point.

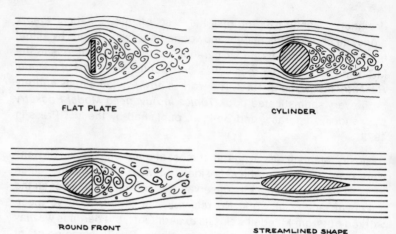

FLAT PLATE

CYLINDER

ROUND FRONT

STREAMLINED SHAPE

Figure 8–11: Typical airflow patterns around objects of various shapes.

So far, we have been talking about some areas of aerodynamics without delving deeply into the subject itself. Basically, aerodynamics is a branch of fluid dynamics that is concerned with only one fairly complex, composite gas made up of oxygen, nitrogen, carbon dioxide, and a few other gases. This special gas is the earth's atmosphere, the air. Aerodynamics is involved with the way air flows around various shapes and with the forces created in the process.

Many years ago backyard inventors and "aeroplane" builders learned that some shapes have less drag than others, and that some shapes have more lift force than others. There is a tremendous amount of information available on this subject. Some of it dates back to the Wright Brothers and earlier. For example, some of the research used by James Barrowman in evolving his *CP* calculation method came from work done during World War I on biplanes such as the Sopwith Camel. Thus, our space-age hobby really has historical roots that go back to the days of the Red Baron! All of this data is readily available to model rocketeers, and a great many of them have used it.

Air can exert considerable force, as we have seen in the previous discussions of flight dynamics and stability. Uncontrolled, this aerodynamic force can tear a model rocket to pieces. Properly used, it can stabilize a model rocket. Much of the force generated by air in motion depends upon the shape of the body around which it flows. If you put your hand out of a car window at 55 miles per hour (carefully, please), and if you open your palm broadside to the airflow (at a 90-degree angle of attack, that is), you will feel a definite push against your palm. If you make your hand into a fist, this semispherical shape will have less drag force created by the airflow.

The drag force is basically caused by the fact that the object must push the air molecules out of the way, slide through them, and permit them to close in behind it again with the least amount of disturbance. Figure 8-11 shows how air can be visualized as flowing around some objects of different shapes. Thus, the shape of an object and the pattern of the airflow around it have a tremendous effect on the drag force created.

There are several forms of drag that are of interest to model rocketeers. They may be summarized as follows:
1. Friction drag
2. Pressure drag
3. Interference drag
4. Parasite drag
5. Induced drag

Let's explore each of these types in turn, because drag is very important in the flight of model rockets.

Air is made up of molecules and is a mixture of gases. For our purposes in model rocketry the molecules can be considered as homogeneous air molecules rather than as a mixture of molecules of different types. These air molecules are so small that trillions of them would fit on the period at the end of this sentence. You can think of the multitudes of molecules as tiny Ping-Pong balls, separated from each other by very small distances. At the earth's surface under normal conditions of pressure and temperature, an average air molecule can travel only 0.000002419 foot before hitting another air molecule. So things are crowded, and the Ping-Pong molecules are hitting each other all the time, creating a gross effect we call pressure. When the molecules slide over the surface of a nose, body tube, or fin, friction is created between the surface and themselves. Actually, even though the surface seems to be shiny and smooth to the eye, it is full of microscopic hills and valleys. *Friction drag* is caused by the air molecules bumping into hills, rebounding off valley walls, and bumping into each other as well. The rougher the surface, the more numerous are the microscopic hills and valleys for the air molecules to hit, and the greater the friction drag.

Actually, the airflow next to the surface exhibits some rather unusual and unsuspected activity. Right on the surface the friction and viscosity of the air slows the first layer of air molecules almost to a

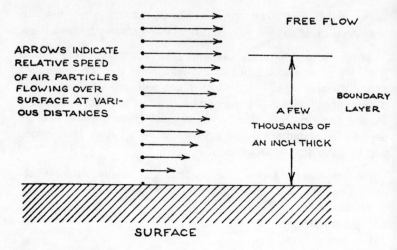

Figure 8–12: The boundary layer.

standstill. The next layer slides and slips over the first layer at a little higher speed. So it goes, with each successive layer sliding faster over the layer below until the full free stream air velocity is reached. This fluid flow phenomenon that takes place close to a surface is called the *boundary layer*. It is shown diagrammatically in Figure 8-12.

All objects moving with respect to the air have a boundary layer. The thickness of the boundary layer, provided that the layers are slipping over one another in an orderly fashion as shown in Figure 8-12, varies with the size of the model and the speed of the airflow. For an average model rocket traveling at about 250 feet per second, the boundary layer is only about 1/1000 inch thick (0.001 inch).

You can actually experience a boundary layer on a beach. If you lie down on the sand, you will be in the boundary layer of the beach surface. You will feel less wind speed there than when you are standing, and you will be able to keep warmer on chilly days by the ocean!

The boundary layer can be laminar, as shown in Figure 8-12. Or it can be turbulent, as shown in Figure 8-13. When a boundary layer transitions, or changes from laminar to turbulent, the layers of air molecules no longer slip easily over one another, but swirl and eddy about. This makes the boundary layer become thicker, even though it is still attached to and flowing over the surface.

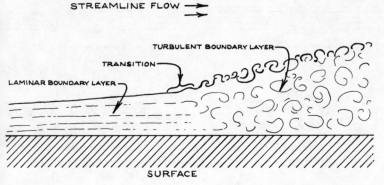

Figure 8–13: Boundary layer transition from laminar to turbulent.

You can see laminar and turbulent boundary layers very plainly by playing around a bit with a kitchen water faucet. Open the faucet slowly and carefully until the water streams out in a smooth, clear fashion. Then open the faucet carefully some more, and the stream will suddenly break into turbulence, sometimes several inches from the spout. If you continue to open the faucet and increase the flow speed, the entire stream will become turbulent. Water is a fluid, like air, and obeys all the same rules even though it has a higher density and viscosity and will not decrease its volume under pressure.

Even with the smoothest surfaces, a boundary layer becomes turbulent at some distance from the front of a model rocket. This distance is a function of the size, shape, finish, and speed of the model. Because of the swirling eddies of a turbulent boundary layer, friction drag is much higher in it than in a laminar boundary layer.

A very small irregularity on the surface will cause transition from laminar to turbulent boundary layer flow. For most model rockets flying at 250 feet per second, a protuberance only 0.0003 inch high will cause transition, or will trip the boundary layer from laminar to turbulent.

There is some data to indicate that the boundary layer on a model rocket usually transitions, or trips, on the nose section itself or close to it. Some modelers theorize that it is best to deliberately trip the boundary layer at the nose-body joint because it is going to trip there anyway because of the joint. Therefore, they make the nose smooth but give the body tube a glossy-mirror finish to reduce friction drag. There is some evidence to indicate that they are right. Experiments

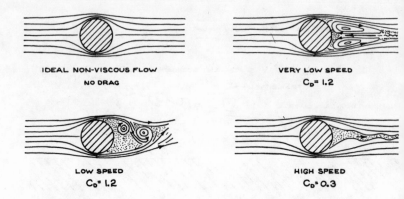

IDEAL NON-VISCOUS FLOW
NO DRAG

VERY LOW SPEED
$C_D = 1.2$

LOW SPEED
$C_D = 1.2$

HIGH SPEED
$C_D = 0.3$

Figure 8–14: Airflow around a cylinder under various conditions.

conducted with a model rocket in a wind tunnel by Mark Mercer, a young rocketeer from Bethesda, Maryland, were reported several years ago by Centuri Engineering Company and indicated a 24% increase in drag coefficient between a rough nose and a smooth nose.

To overcome the effects of friction drag, it is important that the entire surface of the model be as smooth and glossy as possible.

The impact of tiny air molecules on the surface of an object, such as your hand sticking out of the car window, creates a drag force known as *pressure drag*. Obviously this pressure on the front end of an object moving through the air is caused by the impact of air molecules on the object. But it is also possible to have *negative* pressure, a region of fewer air molecule impacts than the surrounding air. (Remember that the body is immersed in the earth's atmosphere, which produces a constant, static, ambient environmental pressure of 14.7 pounds per square inch at sea level.) The region behind your hand in the auto airstream has negative pressure, a partial vacuum.

If there were no boundary layer and if the air had no viscous characteristics, making it like very thin syrup, there would be no pressure drag whatsoever. This condition is shown in Figure 8-14. The air would move smoothly apart in front of the object, allow the body to pass through, and then close in quietly and completely behind the object with nary a ripple. This is the case only for very small objects such as raindrops moving at very low speeds through the air.

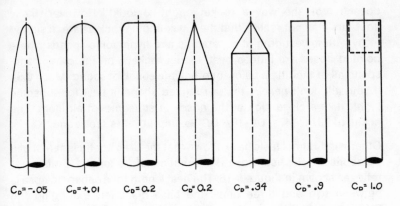

$C_D = -.05$ $C_D = +.01$ $C_D = 0.2$ $C_D = 0.2$ $C_D = .34$ $C_D = .9$ $C_D = 1.0$

Figure 8–15: Pressure drag coefficients of various nose shapes.

An object the size of a normal model rocket moving through the air at 250 feet per second simply does not give the air time enough to slip out of the way and close in again behind it. In other words, the air must be shoved aside by the nose, an action that creates pressure drag and the boundary layer. If you've done a good design and construction job, the boundary layer thus formed finally breaks away from the model at its blunt rear end and goes into a swirling, eddying motion. Thus a wake is created, just like that behind a ship in the water. The wake is a low pressure area that tries to retard, or slow down, the model, and it must therefore be considered as a form of drag. It is called *base drag* and is part of the total pressure drag of the model.

It is possible to reduce the pressure drag on the front end of a model to practically nothing by shaping the nose properly. Many tests on forebody shapes were made in wind tunnels in the United States and Germany during World War II. The results of these tests are summarized in Figure 8-15. Just rounding the edge of a blunt nose by as little as 10% reduces the pressure drag enormously, as you can see.

This wind tunnel data clearly indicates that the best low-drag nose shape for a model rocket is not the sharply pointed ogive shapes of the supersonic sounding rockets, but the rather blunt-looking, rounded shape of a parabola of revolution, or an ellipse of revolution, which is somewhat easier to make.

The rounded nose of a high-performance model rocket usually causes rookie model rocketeers to shake their heads in disbelief. "It doesn't look the way a rocket ought to look!" However, these doubters may not realize that the big sounding rockets are designed to fly with lowest drag at supersonic and hypersonic speeds where pointed noses are indeed best. Model rockets, on the other hand, spend all of their time below one-half the speed of sound (Mach 0.5). If you still don't believe that a rounded shape is best for subsonic flight, take a close look at the noses of subsonic jet airliners. The noses of 707s, 747s, DC-10s, DC-9s, and L-1011s are round.

Other wind tunnel tests have shown that the drag coefficient (C_d) of a parabolic or elliptical nose is a function of its length-to-diameter ratio, as shown in Figure 8-16. The best length for a subsonic nose is between two and three times the base diameter. Extending the length beyond this ratio does not lower the pressure drag appreciably and may actually increase the total drag due to the greater surface area.

Usually there is little that a model rocket designer can do to eliminate the base drag of a model rocket. Most models have a body diameter that is not very much larger than the model rocket motor itself. And the rear, or nozzle, end of a model rocket motor is very blunt indeed! Base drag is reduced during powered flight because the motor is thrusting, pumping billions of gas molecules into the base area. This is often evident after a flight by the slight staining of the base area by the yellowish-brown solid particles in the exhaust plume. However, the base drag jumps during coasting phase. To date there is very little data to indicate what the base drag really is

Figure 8–16: Pressure drag coefficient of a parabolic nose shape in relation to its length-to-diameter ratio.

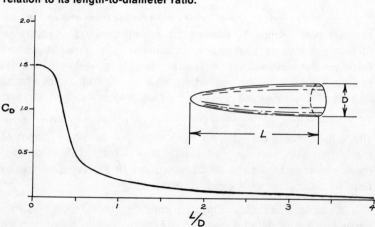

during coasting flight. It is not the "pure" base drag experienced by blunt-ended bodies in wind tunnels because the motor's delay charge is putting out gas, which tends to relieve the low pressure area in the wake. We lack firm experimentally verified numbers on this subject. It remains a good research project for serious model rocketeers.

Some model rockets have body diameters that are greater than the diameter of their motors, perhaps because they must carry large payloads. The base drag of such large-diameter models can be very high.

The classic technical paper on model rocket drag was written in 1967 by Dr. Gerald M. Gregorek of the Aeronautics and Astronautics Department of Ohio State University. Entitled "A Critical Examination of Model Rocket Drag for Use with Maximum Altitude Prediction Charts," Dr. Gregorek's report pointed out that the base drag of a large-bodied model is equal to the base drag of that model shape with a base the same diameter as the motor times the factor:

$$\left(\frac{\text{diameter of the base}}{\text{diameter of the body}}\right)^3$$

One of Dr. Gregorek's students, a model rocketeer named George M. Pantalos, gives a dramatic example of this factor at work. If you have a model with a body diameter of 1.84 inches propelled by a motor with a diameter of 0.75 inch, a model with a base diameter of 0.75 inch would have a base drag equal to:

$$\left(\frac{0.75}{1.84}\right)^3 = 0.41^3 = 0.0677$$

or only 7% of the base drag of the large-diameter tube! In other words, the large model rocket has 93% more base drag than the smaller rocket!

To reduce the base drag of large models, designers reduce the base area by tapering, or making a boattail, as shown in Figure 8-17. According to Dr. Gregorek and George Pantalos, the best taper angle lies between 5 and 10 degrees with 6 degrees being the optimum.

Most model rocketeers make their boattails from stiff card stock paper, designing them in accordance with the principles shown in

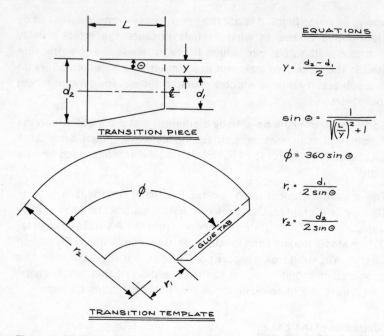

$$y = \frac{d_2 - d_1}{2}$$

$$\sin \Theta = \frac{1}{\sqrt{\left(\frac{L}{y}\right)^2 + 1}}$$

$$\phi = 360 \sin \Theta$$

$$r_1 = \frac{d_1}{2 \sin \Theta}$$

$$r_2 = \frac{d_2}{2 \sin \Theta}$$

Figure 8–17: How to lay out a conical boattail, or transition, of any shape, size, or angle from a flat sheet of material.

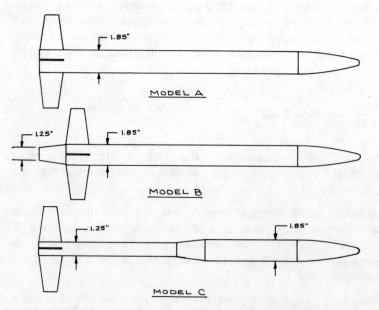

Figure 8–18: The three basic model rocket shapes tested by Pantalos.

Figure 8-17, which shows you how to make a tapered boattail from a flat piece of paper. You can use this data to make a transition piece, or boattail, with angle and lengths of your choice, of course.

Additional work by Pantalos in the Ohio State wind tunnel in 1973 showed that there is an optimum place to put the boattail, or transition. Pantalos tested the three basic model shapes shown in Figure 8-18. Model A is a typical large-diameter model with constant large diameter and no boattail. Model B is Model A with a boattail at the rear of the body, reducing the diameter from that of the model to that of the motor. Model C has its diameter reduction, or transition, right behind that section of the model requiring the large tube diameter for payload purposes.

Which model had the least amount of drag when tested in the wind tunnel? If you guess Model C, you are correct. It has the least body surface area, and so the least amount of friction drag. The base drag of both Model B and Model C is reduced to the absolute minimum through use of the boattails. But Model C has less friction drag. See how these forms of drag interact with one another and how designers must work technical trade-offs?

Interference drag is caused by the interruption of the boundary layer airflow over the body and the fins by the junction between the body and fins. That is, it is caused by the interference of flow between the two surfaces. Technically, it could be considered part of pressure drag. However, the question is raised: Which model will have the least amount of interference drag—one with three fins in triform configuration or one with four fins in cruciform configuration? You should be able to answer that! It is obvious that the model rocket with three fins has 25% less interference drag because there are only three body-fin joints instead of four. This is why most high-performance model rockets have only three fins.

And that leads to an interesting little story. The first three-finned rocket was the historic WAC Corporal, designed at the Jet Propulsion Laboratory of the California Institute of Technology in 1945. Before 1945 all rockets and bombs had four fins, and some aerodynamicists did not believe that the WAC Corporal would be stable in flight. Dr. Frank Malina, the rocket's designer, quietly pointed out that arrows with three fletching feathers had been flying in a stable condition for centuries. Often it pays to bridge the gulf between two apparently unrelated fields of human endeavor!

Interference drag can be reduced by use of fillets, as shown diagrammatically in Figure 8-19. The optimum fillet radius should be

165

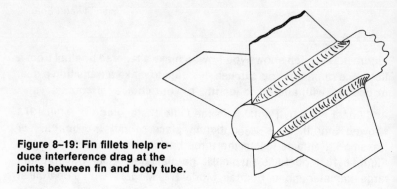

Figure 8-19: Fin fillets help reduce interference drag at the joints between fin and body tube.

4% to 8% of the fin root chord, the dimension of the fin where it is glued to the body. A fillet that is too large will increase the surface area and, hence, the friction drag. A satisfactory fin fillet for most small model rockets can be made with a bead of glue.

Another method of reducing interference drag is suggested by Dr. Gregorek and is confirmed by wind tunnel tests: Move the fins forward of the aft end of the body tube by a distance of about one body diameter. (But be sure to recheck your *CP* calculations!) This slight forward movement of the fins permits the airflow to smooth out on the body surface behind the fins before entering the turbulence of the base drag.

Parasite drag is caused by anything that sticks out from the body to interrupt the smooth flowing of the boundary layer over the model. A major source of parasite drag on a model rocket (which is an extraordinarily clean aerodynamic shape) is a launch lug that is glued to the body tube about halfway along the model. On an otherwise clean, streamlined, well-made model rocket, the launch lug parasite drag can amount to as much as 35% of the total drag of the entire model! Obviously, the way to reduce this drag is to eliminate the launch lug by using a tower launcher. But a suitable compromise is to locate the launch lug in the fin-body joint of one of the fins. This can reduce its parasite drag to only about 20% of the total drag of the model. The effort of relocating a launch lug to reduce drag by 15% is certainly worthwhile!

Induced drag is drag caused by lift, or drag *induced* by the lifting characteristics of a surface such as the model's fins. A body tube and nose combination does not generate enough lift to really matter; in fact, Barrowman conveniently and correctly ignores any nose-body lift in his *CP* calculation method because aerodynamic studies on World War I biplanes indicated that such lift was negligible. The major lift-producing parts of a model rocket are its fins. They do not produce lift as you would normally consider it—lift *against* the force

166

of gravity. They produce lift in the true and classic definition of the term—an aerodynamic force perpendicular to the surface. In model rocketry this lift is part of the aerodynamic stabilizing and restoring force that we discussed earlier in this chapter.

Drag due to lift, or induced drag, is produced by any aerodynamic surface that generates lift. The reason for this is perhaps best understood by looking at Figure 8-4. As you can see, any surface generates drag as well as lift. But some produce less induced drag than others because of two factors—their airfoil shape and their planform shape.

Lift is basically generated by a surface because there is high pressure on the surface exposed most directly to the oncoming airflow and low pressure on the other side. Even a flat plate generates lift. But specially designed shapes called airfoils generate lift better.

The most important source of induced drag, however, is the airflow around the tip of the fin. If a fin is at an angle of attack, there is high pressure on one side of it and low pressure on the other. The high pressure tends to spill over to relieve the low pressure side, and it spills over the fin tip. This produces a spanwise motion of air near the tip. Because the air is also flowing backward over the fin and tip, the result is a corkscrew, or vortex, of air that is shed from the tip of the fin. It takes energy to create and maintain this vortex at each fin tip, and this energy requirement shows up as induced drag. The smaller and weaker the vortex at the fin tip, the less the induced drag.

To this, I hear a young, loud voice say, "Well, just put another surface over the tip of the fin, making a tip plate to prevent the air

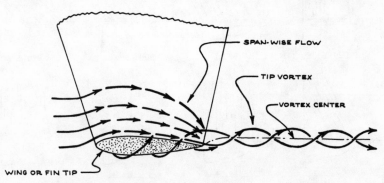

Figure 8-20: Generation of the tip vortex of a wing or fin generating lift.

167

Stability and Shapes

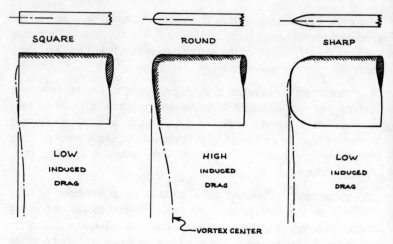

Figure 8–21: Induced drag of three typical fin tip shapes.

from spilling over in the first place." But this just adds more interference drag at the joint between the fin tip and the tip plate!

The best way to reduce the tip vortex is to shape the tips of the fins properly. Induced drag of several styles of fin tips is shown in Figure 8-21. Again, this data comes from wind tunnel tests made to determine the best shapes for airplane wings. After all, a model rocket fin is nothing more than a small wing. You might not suspect

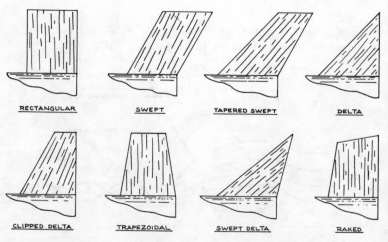

Figure 8–22: Common fin shapes used in model rocketry.

that an absolutely square fin tip would have the lowest induced drag of all shapes tested. And a sharp tip also has the least drag. Both shapes in combination hinder the spillover of the high pressure air to relieve the low pressure on the other side.

The planform shape of the fin has a great deal to do with its induced drag, too. Several common model rocket fin shapes are shown in Figure 8-22. The swept-back fins that look so good on a model rocket are actually the worst when it comes to induced drag.

The two best planform shapes are the clipped delta and the elliptical. The difference in induced drag between these two shapes is less than 1% in favor of the elliptical. But the clipped delta is easier to make. Therefore, I have been personally partial to the clipped-delta planform for a long time and have designed many high-performance, record-setting model rockets using this fin planform.

The basic design parameters for the clipped-delta fin planform are shown in Figure 8-23. It is designed strictly on the basis of body diameter. The fin ends up slightly oversized for most model rockets, so check the *CP* calculations. If you have to cut down the fin area, reduce the root chord and the tip chord, not the span. If you have to increase the fin area, increase the fin span.

Now, how do you put all of this drag and shape information together into a low-drag, high-performance model rocket design? A typical idealized low-drag model rocket design is shown in Figure 8-24. It has an elliptical nose with a length-to-diameter ratio of three. It has three fins with clipped-delta planforms. The launch lug is nestled in the root of one of the well-filleted fin-body joints. The aft end of the tube could not be boattailed because the motor fits right into the

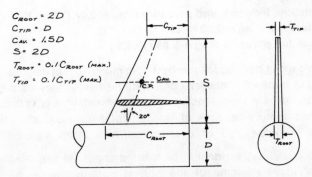

Figure 8–23: How to lay out a low-drag clipped-delta fin planform.

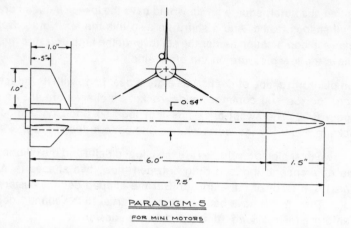

Figure 8–24: A typical low-drag model rocket design.

body tube. So the aft end of the body tube is slightly rounded with fine sandpaper. It is surprising how much the base drag can be reduced simply by rounding the end of the body tube! The model has a very fine shiny, mirrorlike surface with all balsa grain and tube joints filled.

To prove that this idealized model is something that can be built, Figure 8-25 shows some of my low-drag competition designs that have flown and won in national competitions.

The drag coefficient for an idealized model rocket such as Figure 8-24 is approximately 0.4 at a velocity of 200 feet per second. The drag coefficients of the various parts have been calculated by Dr. Gregorek and George Pantalos as follows:

Nose and body pressure and friction drag = 0.2 (50%)
Base drag = 0.06 (15%)
Fin friction, induced, and pressure drag = 0.07 (17.5%)
Interference drag = 0.02 (5%)
Launch lug parasite drag = 0.05 (12.5%)

In typical engineering fashion optimum model rocket designs are the result of trade-offs between conflicting elements. They are technical compromises. A model rocket designer's handiwork is as distinctive as his signature because of the particular trade-offs and compromises he makes. You will develop your own design signature, too.

There is still a great deal to be learned about the aerodynamics of model rockets. Some of the hottest controversy in the hobby revolves around aerodynamics. All of the data hasn't come in yet.

Figure 8–25: Some of the author's low-drag, high-performance model rocket designs. Fourth from right is P-Chuter, which took the bronze medal in the parachute duration category at the First World Championships for Space Models.

There is a lot of guessing, a lot of theory unsupported by experimental data, and a lot of data that has not been correlated into meaningful design rules. This indicates that there are many areas that are ripe for original research, the sort of study that does not require an operating knowledge of advanced mathematics, a supply of expensive equipment, and years of expensive training.

The aerodynamic research area of model rocketry is full of fun and games. Just when you think you've discovered the answer, some other model rocketeer comes along with data that say otherwise— and the controversy is under way. Don't believe that controversy is bad for the hobby. At one early space travel conference the famed Dr. Theodore von Karman was asked to sum up the meeting. He stood and said, "A very fine meeting. Very well-organized. Very well-run. Excellent papers on very pertinent subjects. Good presentations. Solid data. But no arguments! *No arguments*, gentlemen! How in the world can you possibly have progress without controversy?"

So, if you have the data to support your opinions, argue away. In the process you will have fun and learn a lot more. And someday, perhaps, you will graduate to designing the big ones.

171

Multistaged Models

So far, we have considered only single-staged model rockets. There are two main reasons for this. First, nearly everything that applies to single-staged model rockets also applies to multistaged model rockets—and sometimes even more so. Second, unless you understand the principles that we have discussed about single-staged models, you will have considerable difficulty with multistagers.

A multistaged model rocket permits you to increase the performance by increasing total impulse while at the same time greatly decreasing burnout weight. A multistaged model, technically, is one in which two or more motors operate in flight with the expended motors and their airframes being dropped off in flight after they have reached burnout.

With a single-staged model you can increase the burnout velocity by decreasing the weight until the model is at the optimum weight indicated by the Malewicki altitude charts in Centuri TIR-100. You can also accomplish this by super-careful streamlining and drag reduction techniques. And you can increase the total impulse of the motor.

However, there are limits to each of these methods beyond which you cannot go for a number of reasons. Multistaging offers a method of increasing burnout velocity without some of the disadvantages and limits imposed on single-staged models—but with some nasty little tricks of its own that you must watch for!

Figure 9–1: The performance of a model rocket can be increased by utilizing the principles of multistaging used in space rockets.

The simplest form of staging is *series staging*, shown in Figure 9-2. Essentially, the payload of the lower stage is itself a model rocket—the upper stage. A motor and its enclosing airframe—body tube, fins, and motor mount—make up the lower, or first, stage, often referred to as a booster. The upper stage is a complete airframe, a single-staged model rocket. Ignition of the lower stage accelerates the entire model into the air. At lower stage burnout the lower stage booster assembly separates from the upper stage, and the motor of the upper stage is ignited. As shown in Figure 9-3, this effectively

173

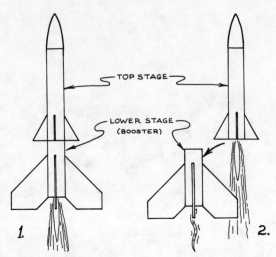

Figure 9–2: Series staging.

increases the total impulse of the upper stage. Since the peak altitude is a function of the square of the burnout velocity of the final stage, and since the burnout velocity of the lower stage is added to that of the upper stage, you can quickly see that the peak altitude of a multistaged model rocket can be very high indeed.

However, aerodynamic drag forces can also be very high.

In addition to increasing the total impulse of the model without requiring extensive redesigning and rebuilding, series staging also decreases the burnout weight of the upper stage because the

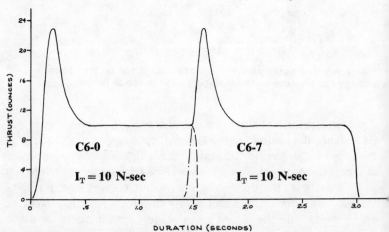

Figure 9–3: Thrust-time curves of series-staged C6-0 and C6-7 motors. Combined total impulse is about 20 N-sec.

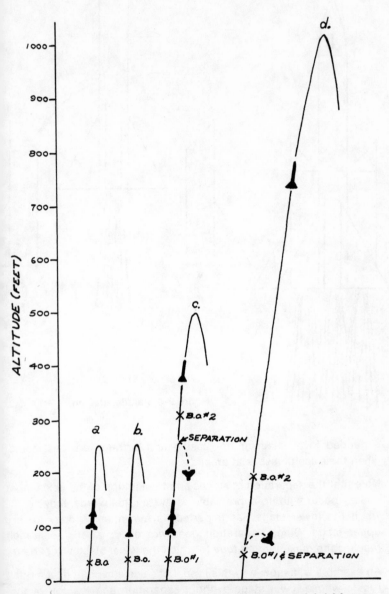

Figure 9–4: Staging at various points in the flight of a model. (a) Lower stage only ignites. (b) Top stage only flies. (c) Model is staged at lower stage peak altitude. (d) Model is staged at lower stage burnout point (B.O.) and maximum stage velocity.

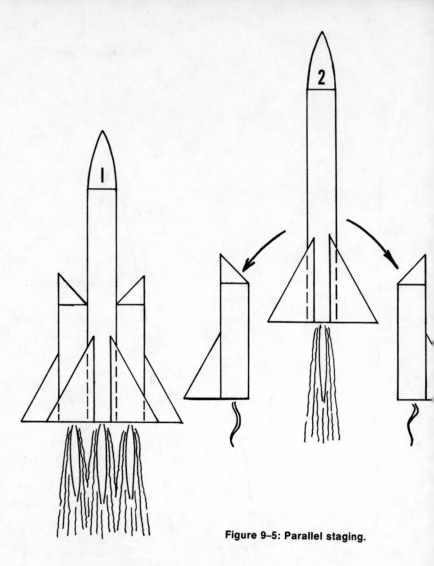

Figure 9–5: Parallel staging.

expended lower stage motor casing and airframe are cast aside
when their usefulness is at an end.

Note that the separation of stages and the ignition of the next stage
usually occur with no appreciable delay. In other words, they occur
when the lower stage has imparted its maximum *velocity* to the
upper stage. Stage separation does *not* occur when the model
reaches the maximum *altitude* to which the lower stage will carry it.

An example will show why this is so. Let's use a hypothetical model
rocket whose lower stage, without separation and ignition of the
upper stage, can carry the entire model to a burnout velocity of 100
feet per second and a peak altitude of 250 feet, as shown in Figure
9-4. Suppose that the upper stage all by itself can achieve a burnout

velocity of 100 feet per second and a peak altitude of 250 feet.

If the staging took place at the peak altitude of the lower stage, the peak altitudes would add together, giving a total peak altitude of 500 feet for the combination staged in this fashion. However, if the upper stage is separated and ignited at the maximum velocity of the lower stage at lower stage burnout, the 100 feet per second of the lower stage is added to the 100 feet per second of the upper stage, giving a burnout velocity of 200 feet per second for the staged combination. Peak altitude is therefore four times what it would be at a burnout velocity of 100 feet per second, and the upper stage peaks out at 1,000 feet.

In other words, a model rocket staged at the maximum velocity of the booster will go roughly twice as high as one staged at the maximum altitude of the booster—neglecting aerodynamic drag, of course.

Another type of staging is called *parallel staging*, because the stages operate in parallel rather than in series. As shown in Figure 9-5, the parallel-staged model rocket leaves the launch pad with all motors of all stages operating. The staged motors/airframes are quite properly called boosters, and the upper stage is the core, or the sustainer.

The booster motors are selected to have less duration than the sustainer motor. Perhaps the sustainer motor is a Type C6 while the booster motors are Type B6 or B14. At booster motor burnout the booster pods are separated by air drag or other techniques, leaving the sustainer motor still thrusting and accelerating the model.

Parallel staging provides very high thrust and acceleration at lift-off. It also eliminates the problem of the air start of series-staged upper stage motors. However, ignition of multiple motors, or clustered motors, is not a simple, easy task in spite of modern ignition techniques and equipment.

For many years it was believed that parallel staging was not practical for model rockets because of the many difficulties involved. Therefore, very little work was done on it although we modelers knew it would be possible. Pat Artis, of Ironton, Ohio, perfected parallel-staged model rockets and demonstrated them at the Seventh National Model Rocket Championships in 1965. By 1966 Artis had refined his designs to carry the booster pods up near the nose of the model, thereby bringing the *CG* of the model well forward during the critical boost phase and eliminating the need for large fins to ensure stability during boost phase. Artis dramatically showed that

Figure 9–6: Pat Artis (left) describes his parallel-staged model rocket to Cal Weiss of NASA at NARAM-8. Note forward mounting of booster pods.

parallel-staged models had superior performance for payload-carrying because of the high lift-off acceleration.

Still, series-staged model rockets remain the most common staged models flown. Over the years I have seen successive waves of popularity for series-staged models, then parallel-staged models. However, the popular series staging has long been subjected to serious problems with the air start of the upper stages. Air-start ignition is a knotty problem that has not yet been fully solved, although others as well as myself are working on it at the time of this writing.

Ignition of the first stage motor in a series-staged model is accomplished electrically in an identical manner to that of a single-staged model rocket.

Figure 9-7 shows in cross section a typical two-staged series-staged model rocket at various times in its flight. Drawing A shows the model shortly after lift-off; the propellant in the lower stage motor is burning forward. Note that the lower stage airframe is nothing more than a hollow body tube with motor mounts installed and fins attached. The body tubes of the two stages are simply butted together. Thrust of the lower stage motor is transferred to the upper stage through the lower stage motor mount, lower stage body tube, and the butt-joint between the stages. The lower stage motor is held firmly in the airframe by a metal motor clip or by friction-fit with a motor mount ring at the nozzle end of the motor—for a reason that will become clear in a moment.

In Drawing B the lower stage motor has almost reached burnout. There is only a thin disc of propellant left in the lower stage motor. Separation and staging are only a split second away.

In Drawing C the hot combustion gases under pressure have broken through the thin disc of propellant and are sending hot chunks of burning lower stage motor propellant forward. These pass up into the nozzle of the upper stage motor, igniting it. The jet from the upper stage motor pressurizes the lower stage body tube and blows it off the model. The rear-positioned motor ring in the lower stage airframe prevents the lower stage motor casing from being ejected from the lower stage airframe during this process.

The upper stage then takes off on its own, leaving the lower stage to tumble back to the ground. Lower stages are usually designed to be

Figure 9–7: Cross section of a multistaged series-staged model rocket at various progressive points of lower stage operation and at separation.

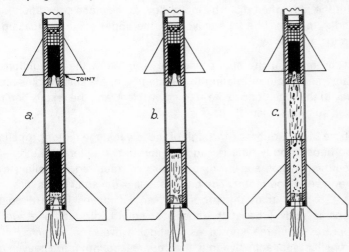

deliberately unstable all by themselves, and this takes some tricky figuring of the Barrowman Method!

However, the biggest problem here lies in ensuring the ignition of the upper stage motor. The problem manifests itself as follows: The model comes up to stage separation, and the blow-through feature separates the stages before the hot pieces of burning lower stage propellant have had a chance to get up into the upper stage nozzle to ignite it.

Various techniques have been developed to delay the separation of the lower stage for a fraction of a second until the upper stage has positive ignition. If you don't get upper stage ignition, it separates, but there is nothing to eject the recovery device because the upper stage motor isn't working. The model coasts upward to apogee and comes right down in a very speedy manner.

The most common delaying method is to design the model so that the lower stage airframe is short and both motor casings butt together. The motor casings are secured together with a single wrap of cellophane tape. This keeps them together during staging until the ignition of the upper stage literally blows the casings apart. The drawback of this method is the requirement to butt the motors together. This can result in short, underdamped series-staged models that require excessively large fin areas to achieve static stability, often at some expense to dynamic stability.

An alternative to the taped-motor method is to tape the airframe stages together. This permits lengthening the lower stage airframe. But the tape often disturbs the airflow at the stage separation joint, and sometimes the tape stays with the upper stage, increasing its drag significantly.

At the moment the best procedure is to follow the instructions of each manufacturer relating to the staging of his kits. Work is under way to find solutions to the staging problem, and the results begin to look something like this:

At the instant of break-through of the lower stage motor propellant, the booster body tube is suddenly full of two different things—hot combustion gas and burning chunks of fractured propellant from the lower stage motor. The hot gas is less dense and has less total heat content than the propellant, and it moves forward up the booster body tube at the speed of sound, producing a sudden overpressure area known as a shock wave. The hot chunks of propellant that actually ignite the upper stage motor are heavier and therefore move slower. By the time they get to the upper end of the booster body tube, the shock wave may already have blown the stage off the model.

It appears that the major problem is to keep the stages together long enough for the hot particles to reach the upper stage nozzle and propellant. Taping motors and stages together accomplishes this, preventing the shock wave from blowing the stages apart for a fraction of a second until ignition has taken place. But there must be better methods.

You probably have noticed that lower stage model rocket motors for both parallel- and series-staged models are somewhat special. They have no time delay and no ejection charge. Their NAR type numbers end in a dash-zero—B6-0 or C6-0, for example. This lack allows for staging and separation at burnout of the lower stage motor.

Do not—I repeat, do not—use a standard motor with a time delay and ejection charge in a lower stage. The model will lift off normally, go through lower stage powered flight, and then start to coast. Because a staged model is heavier at lift-off due to its additional stage airframe and motor, the model will coast upward and arc over through apogee. Because of the time delay, the lower stage will then separate and ignite the upper stage when the model is pointed downward! Under thrust assisted by gravity, and with considerable initial downward velocity, the upper stage will drive toward the ground with frightening speed. There is no time for the recovery device to eject, nor would it stay in one piece if it were activated. The upper stage usually impacts under thrust. Very little usually remains of such reentry models, and they are exceedingly dangerous.

It is a good safety rule to always check carefully that you have installed a dash-zero motor in all lower stages.

Although each lower stage motor should be a dash-zero type, the motor in the final upper stage must have a time delay and ejection charge to deploy the recovery device. Because of the higher final burnout velocity of staged models, the upper stage coasts for a longer period of time before all its momentum is converted to altitude. Therefore, upper stage motors must have a longer time delay to prevent popping the recovery device before peak altitude is reached. Long-delay motors are made and specified for upper stages. When in doubt, however, use a shorter time delay for initial flights to prevent cliff-hangers where the model goes over peak altitude and opens on the way down.

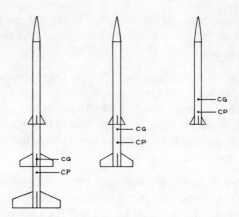

Figure 9–8: Stability checks on a three-staged model rocket.

As stated earlier, multistaged models weigh more at lift-off than single-staged models because of the additional weight of their lower stage motors and airframes. Therefore, with normal motors, multi-staged models lift off slower and accelerate slower during boost. Because they are launched at lower velocities, good stability is important if a safe, predictable flight is to be achieved.

Multistaged models follow the same ground rules for stability as single-staged models. All of the stability requirements must be met. For a two-staged series-staged model the *CP* and *CG* of the two stages together (in launch configuration) must be in a stable relationship. The *CG* and *CP* of the upper stage alone must be properly located. Thus, stability checks must be made for every flight combination of a multistaged model—in its various staged configu-rations and its final top-stage configuration. For a three-staged series-staged model three stability checks must be made—for all three stages together, for the top two stages together, and for the top stage all by itself. See Figure 9-8. Although you may occasionally get away with flying a new single-staged model without a stability check, you should *always* make stability checks for multistaged models. At the very least, subject the completed model to a swing test.

Because of the lower lift-off velocities and larger fin areas of multistaged models, they are very susceptible to weathercocking. Therefore, they should be launched only in calm conditions or with winds of less than 5 miles per hour.

Booster stages are made with the same techniques as those used in building upper stages and single-staged models. In essence, the upper stages of a multistager become the "nose" of the lower stage. Recovery devices are not normally installed in lower stages (although I have done it) because of the need for the blow-through to ignite the next stage. A lower stage airframe, however, must be built in a more rugged manner than a simple single-stager. Although it is designed to tumble-recover, it usually has very large fins, which are easily broken, even on a normal soft landing. Double-glue joints must be used throughout. Fin joints should be reinforced with tissue fillets.

Because lower stages do not deploy a brightly colored recovery device like a single-stager or an upper stage, they are more difficult to locate after they have landed. Therefore, they should be painted a very bright fluorescent color so that they can be located on the ground.

The couplings between stages should be designed and built so that the model will not jackknife in the air at the joint. Most kit models have this sort of solid joint, and there are numerous different kinds. Basically, the joint should permit lengthwise stage separation but prevent wobbling in the pitch and yaw axes. It is quite frightening during multistaged boost to have the model jackknife in midair and come apart. You're never quite sure where the top stage is going to go if it manages to stay together long enough to achieve an air start. It is a very expensive form of Russian roulette.

Launching lugs may be placed almost anywhere on a multistager, but here is a tip: Put the launch lug in a fin-body joint fillet only on the upper stage. This permits you to fly the model as a single-stager, too.

To achieve maximum possible stability, do not align the fins of various stages in a fore-and-aft direction. This puts the lower stage fins in the wake, downwash, and vortex pattern of the upper stage fins and greatly reduces their effectiveness—and you need all the effectiveness you can get from lower stage fins. Instead, interdigitate the fins. Put them out of line with each other, as shown in Figure 9-9.

Upper stages can achieve some pretty high burnout velocities, sometimes well in excess of half the speed of sound (Mach 0.5). They should be well-built with exceptional care devoted to drag reduction if maximum performance is desired. They should also be built strongly because the drag force on an upper stage can exceed the weight of the model by a factor of two or more in some cases. I have seen upper stages come completely apart after separating, leaving fins and other parts all over the sky. They have reached the legendary speed of balsa.

We have been speaking generally of two-staged models with a few references to three-stagers. However, the principles apply to three-stage models, too. The safety rules of the National Association of Rocketry and the Code for Unmanned Rockets of the National Fire Protection Association have limited multistaged models to a maximum of three stages.

The reason for this limitation is that the reliability of series-staged model rockets appears to decrease according to the inverse square of the number of stages. Thus, a two-staged model is roughly one-fourth as reliable as a single-staged model, while a three-stage model is only one-ninth as reliable as a single-stager. This would make a four-stager only one-sixteenth as reliable. As noted earlier, it is difficult to get an air start on a two-staged model, so it is a minor miracle to get one twice with a three-staged model. In addition, with the limited total impulse of model rocket motors, a three-stage model may actually fly to a lower peak altitude than a two-staged model. Strange as this may seem, it is due to several factors. First, a three-stager is heavier, lifts slower, and may have a final burnout velocity that is less than a two-stager. If a high burnout velocity is achieved with a three-stager, drag forces may rob it of most of its altitude during coast.

Besides, a properly designed three-stager that is well-built, correctly launched, and lucky will go completely out of sight. Then what are you going to do? You can no longer see it. You've probably lost it!

To boost a three-stager off the pad and get it going in a stable manner rather than lumbering and wobbling around the sky, use a high-thrust motor in the first stage. Such a motor is the B14-0 or the A10-0T. They provide high thrust that kicks the heavy model into the air properly. Their burning time is very short, but the model will likely be at a higher altitude at stage separation because of a straight, true boost flight. There is always considerable trepidation when flying a three-stager, but less so with a high-thrust motor in the lower stage.

When everything works and the design is right, multistaged model rockets are capable of outstanding performances. But because of the great potential for something to go wrong, all safety precautions must be rigorously followed with great care. Tilt the launch rod a few degrees away from the vertical so that the model will land down-

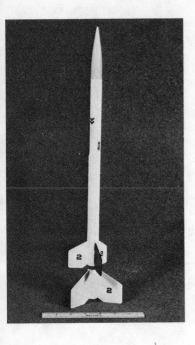

Figure 9–9: This two-staged model illustrates interdigitated fins. The upper stage fins are not aligned with the lower stage fins.

range and away from people and cars if it has a staging failure and doesn't get an air start.

Because of the high-performance and high-altitude capabilities of multistaged model rockets, you should always keep in mind the Federal Aviation Administration's regulations with respect to model rocket flight operations. These are spelled out in detail in Federal Air Regulations Part 101. Model rockets are exempt from FAA control if they weigh less than 16 ounces at lift-off, contain less than 4 ounces of propellant, are made of frangible nonmetallic materials, and are flown in such a manner that they present no hazard to aircraft in flight. Model rockets exceeding these operational limitations must be covered with an FAA Air Traffic Control clearance, which is difficult and time-consuming to obtain. It is easily possible to exceed the FAA limits with multistaged models. So beware and be warned.

Multistaged models are fun and difficult. If anybody tells you that model rocketry is kid stuff and playing with toys, he hasn't built and flown a multistaged model rocket. And there are still a lot of unanswered questions in multistaged technology. It is a good area for research and development activities that will result in greater reliability and better performance.

Recovery Devices

Because recovery devices are required in all model rockets, the hobby is virtually free of any hazard caused by freely falling objects that are large enough and heavy enough to do damage or cause injury when they return to the ground.

The rules of the NAR and the FAI do not permit the jettisoning and free fall of any part of a model rocket, such as the motor casing, unless the falling part tumbles to slow its speed and uses a streamer or other visible surface so that it can be seen. The model itself must be recoverable, capable of more than a single flight, and return to the ground so that no hazard is created. If you do everything right, follow all of the rules, and read the instructions, you have a 99.9999999999% chance of having your model rocket perform in this safe fashion. If something should go wrong, as it occasionally does, the design of the model and the materials used in its construction are intended to reduce the hazard to acceptable levels.

Although we have discussed recovery devices in general terms earlier, we'll now cover them in detail because there are all sorts of little hints, kinks, and tricks about making and using them.

A number of highly successful and reliable recovery devices have been developed. The type of recovery device used in a model rocket depends upon many factors—gross weight of the model, recovery weight of the model, type of payload, size of the model, etc.

All recovery devices in solid propellant model rockets are actuated or deployed by the ejection charge of the model rocket motor. This

Figure 10-1: Model rockets use recovery devices such as a parachute to lower them gently and safely to the ground so that they may be flown again—provided they miss the rocket-eating trees shown in the background.

quick burst of gas from the ejection charge can be made to do a number of things to activate a recovery device. In cold propellant model rockets different systems are used, as we discussed earlier.

Nose-blow recovery

I don't recall when the first nose-blow recovery system was used or who used it. It was definitely in use in the summer of 1958 during the flight testing of simple models in Denver. It may have been first used by some model rocketeer who took the parachute out of his model to save time during flight preparations or to prevent the model from drifting away in a high wind.

Nose-blow is a very simple recovery system that derives from a similar system used in 1946 and thereafter at White Sands Proving Ground for aiding the recovery of German V-2 rockets. Normally, a 3-ton V-2 rocket falling from an altitude of 100 miles would dig a rather large, deep hole in the desert, completely destroying all the instruments as well as the rocket vehicle itself. One day a V-2 happened to come apart on the descending leg of its trajectory and tumbled to the ground. The astounded rocket scientists recovered most of their equipment intact. Thereafter, the flight safety officer merely activated the ring of explosives around the base of the nose cone during the descending leg of the flight. The V-2's streamlining and stability were destroyed by this action, and the separated pieces fell at a much lower terminal velocity.

In model rocketry the same principle lies behind nose-blow recovery, but we do it a bit differently. The nose is tied to the rest of the model by a shock cord. When the nose is separated from the model, the aerodynamics of the model are ruined, and the *CG-CP* relationship is altered so that the model is unstable. With the nose tied to the model by the shock cord, the two parts land together; this eliminates the need to search for each piece separately.

A typical nose-blow model is shown in Figure 10-2.

Although a nose-blow model falls slowly enough to catch with your bare hands, it can still land hard enough to break fins if the landing site is asphalt, concrete, or hardpan. Therefore, most nose-blow models are constructed very strongly. And nose-blow recovery is used mostly for models weighing less than 2 ounces, or 60 grams. It is also used for high-altitude models when you don't want them to drift into the next state. However, occasionally one sees nose-blow used on large models with a high area-to-weight ratio where terminal speed is low. Nose-blow is almost never used when flying a fragile payload.

Streamer recovery

When a model rocketeer wishes to slow his model a bit more than is possible with nose-blow recovery and also wishes to see it more clearly against the sky and on the ground after landing, he adds a streamer to the nose-blow recovery system.

A streamer is a bright-colored strip of plastic film or crepe paper attached to the shock cord or to the nose base. The streamer is usually attached only at one end, although it may also be attached in the middle. Streamer dimensions vary between 1 inch and 2 inches in width and 12 inches to 36 inches in length. Many modelers, myself

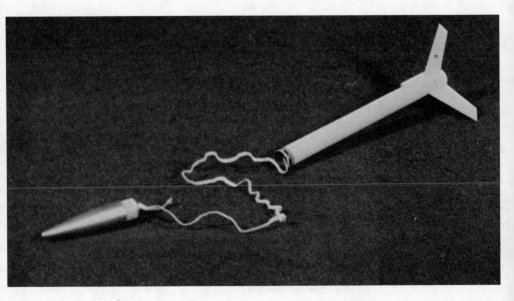

Figure 10-2: A nose-blow model rocket uses the ejection charge of the motor to separate the nose from the body, destroying the stability. The nose is tied to the body by a shock cord.

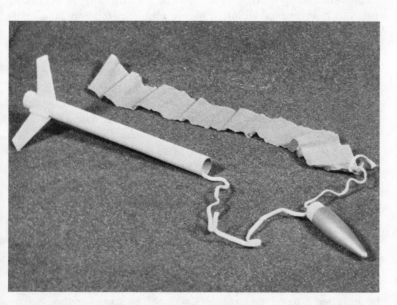

Figure 10-3: A streamer-recovery model rocket. The streamer may be attached by a separate line to the nose base or may be tied to the shock cord near the nose.

included, prefer to use bright-colored crepe paper instead of plastic because it does not get stiff in cold weather, seems easier to fold and roll, does not burn or melt if some ejection gas seeps around the recovery wadding, and offers superior drag characteristics.

Some work done by Trip Barber and others at the Massachusetts Institute of Technology indicates that there is an optimum streamer size for every model design. Generally, the optimum streamer size to obtain the slowest descent rate has a length-to-width ratio of 10; in other words, a 1 inch by 10 inch streamer or a 2 inch by 20 inch streamer. Apparently, according to the tests run by Barber at MIT, one does not gain anything by going to a streamer longer than 10 to 1. A long streamer seems to actually stream. To slow a model down, a streamer must flutter, changing its angle of attack.

Barber's work also indicates that it is possible to match the streamer dimensions to a given model rocket design to achieve the lowest possible descent rate and therefore the longest flight duration time. For this reason, the NAR has placed no limits on streamer size or material and has not attempted to adopt a standard streamer for its Streamer Duration Competition. This class of competition was originally intended as a simplified achieved-altitude contest for beginners and for "fun" meets. However, like other types of competition, it quickly evolved into a highly sophisticated contest with its own strategy and tactics. In spite of this, however, Streamer Duration remains the simplest of all competition categories to fly.

A streamer must be protected from the hot gas of the motor's ejection charge. A plastic streamer can be melted and a crepe streamer can be burned by the ejection charge gas. Protection is simple and easy: Stuff a small wad of flameproof tissue down into the body tube and put the rolled-up streamer in on top of it. The wadding blocks the ejection charge gas from the streamer and also acts as a piston to help eject the streamer. The wadding is ejected from the model after the streamer comes out; it should always be made from flameproof material so that there is absolutely no chance it may be ignited by the ejection charge and fall smoldering to the ground, where it could start a fire in dry grass.

A streamer is packed into a model by first rolling it tightly into a long cylinder. Often a single line of string or thread is used to attach the streamer to a screw eye at the base of the nose. This string can also be wrapped around the streamer cylinder. The streamer should slide easily into the body tube so that it cannot be jammed. It should be capable of being ejected easily. If you wad it in, the ejection charge

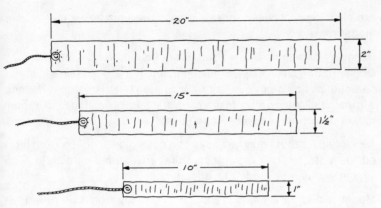

Figure 10-4: The optimum streamer size is a length-to-diameter ratio of 10 to 1, as shown.

won't be able to eject it. Do not push the streamer into the body tube more than about an inch, leaving enough room to install the nose; you want all the weight you can get up front to improve the *CG-CP* relationship.

A streamer recovery device will cause the model to drift in the wind a great deal farther than nose-blow recovery. Often the model hangs up in a tree. In fact, it is a foregone conclusion among experienced model rocketeers that a model is certain to land in the top branches of any tree within sight of the launch area at some time during the day's flying activities.

Parachute recovery

Parachute recovery of rockets is an old and established art. It was first reported in a description of a fireworks demonstration conducted by the Ruggieri brothers in Paris shortly after the French Revolution. It may have been used by the Chinese themselves in early fireworks skyrockets. Dr. Robert H. Goddard successfully recovered many of his rockets with parachutes, and early experiments were made by the American Rocket Society in the 1930s using parachute recovery. Many modern research and sounding rockets regularly use parachutes for the recovery of the payload or the entire vehicle. Some of these rockets even have replaceable motors like model rockets.

Recovery parachutes were used in the first model rockets made by Orville H. Carlisle in 1954. Since then, much has been learned about parachutes, and many new techniques have been developed. In fact, a great deal of serious research work has been done on model rocket parachute recovery techniques.

A simple parachute may be added to a nose-blow model or substituted for a streamer. It is surprising that a workable parachute is so simple and can be made to work so reliably.

Model rocket parachutes are commonly made from thin polyethylene film ranging in thickness from 0.00025 inch to 0.001 inch. Model rocket manufacturers sell parachute kits that include brightly colored parachute canopies of various sizes and shapes. These colored canopies are easier to see against the sky than parachutes made from clear dry cleaner's clothing bags.

Some high-performance competition parachutes are made from a very thin and very strong plastic film called Mylar. Such thin parachutes are also used when the storage space inside the model is very small. Mylar film is available in aluminized form in the United States, and it is superior for parachute canopies because it glints in the sun and can be seen for long distances through heavy haze. Aluminized Mylar parachutes also have a fine "memory"; they "remember" the flat condition in which the plastic film was originally laid down in the factory, and they tend strongly to open up and resume this flat condition even at very low temperatures.

Ordinary polyethylene parachutes get stiff and hard at temperatures below 40° F. Under such conditions they sometimes fail to open, resulting in a recovery device known as a plastic wad; it does not have a very high drag coefficient.

Figure 10-5: A model rocket with a recovery parachute made from polyethylene film and using twine or thread for shroud lines.

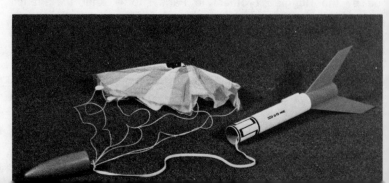

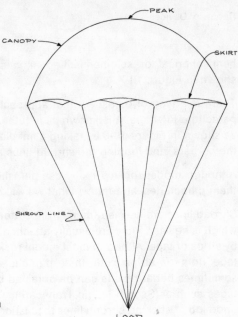

Figure 10-6: The parts of a parachute.

Parachutes used in model rocketry are simple flat sheets of material. Therefore, they are technically not parachutes, but parasheets. A true parachute is not made from a flat piece of material and cannot be laid flat on the ground; it is made from wedge-shaped gores of material that are sewn or fastened together to make a canopy that is

Figure 10-7: A true parachute is made with gores of material joined together to create a hemispherical shape when inflated.

hemispherical or semihemispherical in shape when inflated, as shown in Figure 10-7.

There is a type of parasheet that approaches a true hemispherical parachute in shape. It is known as a gathered parasheet. It is made as shown in Figure 10-8 by taking a flat piece of material, gathering the corners, and looping the shroud lines around each corner.

Although model rockets really use parasheets, it is simpler to call them parachutes, and that is what we will do from here on.

Shroud lines are normally made from cotton thread or carpet thread, which is heavier. They are usually attached to the skirt of the canopy by strips of tape. Although most model rocket manufacturers supply tape dots or strips with their parachute kits for this purpose, sometimes better results can be obtained by using readily available tapes such as Scotch Magic Transparent Tape, which does a very good job of sticking shroud lines to plastic 'chutes; it doesn't pull off as easily as some tape dots or strips will, and it does not lose its holding power with age. Freezer tape works well, too.

To prevent the shroud lines from pulling away underneath the tape, loop them under the tape or tie a knot in them, as shown in Figure 10-9.

You should try to make the shroud lines all the same length. It will result in a more symmetrical parachute.

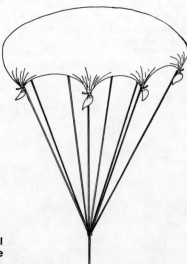

Figure 10-8 : A gathered parasheet is made from a flat piece of material gathered where the shroud lines are attached.

194

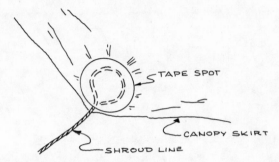

Figure 10-9: Shroud lines are usually looped under the tape that attaches them to the parachute skirt. This prevents them from pulling out from under the tape spots or strips.

There is little data available on optimum shroud line length. However, my experience indicates that the shroud lines should never be shorter than the major dimension of the parachute. Thus, a 12-inch parachute should have shroud lines 12 inches long. Actually, I have been using shroud lines at least 1.5 times the major dimension of the parachute, and I have managed to win a few competitions and lose even more models into thermals where they have disappeared going up.

A good research project would be: How is the sink rate of a parachute affected by the length of the shroud lines? Make up several 'chutes with different length shroud lines. Drop them from known heights indoors in calm air. Time their descent to the ground. Make many drops—the more the better. It sounds like an easy project, and it is. But it needs to be done. Too many model rocketeers become enamored by complex theoretical projects and often neglect researching many everyday problems.

The desired number of shroud lines depends upon the size and shape of your parachute. There are only a few basic shapes, as shown in Figure 10-10. These are: round, square, hexagonal (six-sided), and octagonal (eight-sided). On circular 'chutes six to eight shroud lines appear to be adequate. Most 'chutes have a shroud line at each corner—four for square 'chutes, six for hexagonal 'chutes, and eight for octagonal 'chutes.

Once the shroud lines have been attached to the 'chute, bring the loose ends together in a knot and tie this to the base of the nose. Some modelers use a snap swivel of the sort used by fishermen; this allows them to attach and remove parachutes for different flight conditions.

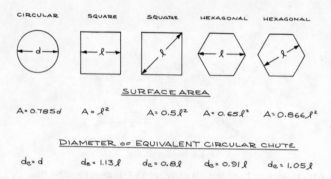

Figure 10-10: Some of the basic parachute shapes and sizes tested by Kratzer *et al* at the University of Maryland.

There is a wide variety of methods of folding and packing parachutes into model rockets. Some people merely stuff the 'chute in atop the wadding. Others develop weird methods of folding. I am still using the 'chute packing method taught to me by Orville H. Carlisle in 1957; it has worked excellently for me over the years. This Carlisle Method is shown in Figure 10-11. Follow the steps shown, and you will have a tightly rolled parachute cylinder that will easily slide down into the body tube and easily be ejected. When the 'chute is ejected, it unrolls very rapidly and deploys quickly. I have heard 'chutes fill so quickly that they pop.

Although I always use wadding except where ejection baffles are built in as part of the model (and they work), I also take one additional step to be absolutely certain that the 'chute does not get scorched or spot-burned by ejection charge gas leaking past the wadding. I wrap the 'chute package in a piece of flameproof wadding tissue. One layer is enough. It peels off quickly upon ejection. And it *always* works. It also prevents the 'chute from getting ripped during ejection.

How big a parachute is required? Well, how long do you want your model to stay aloft? The bigger the parachute, the longer the model will stay in the air. This general observation is based on the fact that a parachute produces a lot of drag. That's its function. Therefore, it must obey the drag equation:

$$D = 0.5\rho V^2 C_d A$$

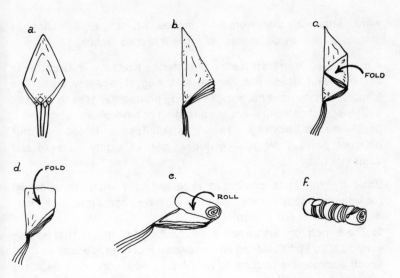

Figure 10-11: The Carlisle Method of folding and packing a parachute.

where this time A = the area of the parachute canopy.

Data from aeronautical engineering texts indicate that the C_d of a hemispherical parachute is about 1.5 while that of a parasheet is roughly one-half of this, or 0.75.

Although model rocketeers knew about this equation and knew that it applied to parachutes, we didn't pay too much attention to it. We went by the rule of thumb that the bigger the 'chute, the longer the flight—and the farther you had to chase the model on a windy day.

Numbers were brought into the picture through a sophisticated series of parachute drop tests carried out in 1970 by Carl Kratzer, Bruce Blackistone, and Larry Lyons and reported by Doug Malewicki. They made a number of timed drops with various types of parachutes and standard loads from a 90-foot platform inside the Cole Fieldhouse of the University of Maryland. This provided them with a controlled environment of still air at a constant temperature. The 'chutes were circular, square, hexagonal, and octagonal. The first test series were made of 0.00075-inch-thick polyethylene 'chutes from Estes Industries, Inc. Sizes were 8 inches, 12 inches, 18 inches, and 24 inches. The second series comprised parasheets made from half-mil (0.0005 inch thick) clear polyethylene made from dry cleaner's clothing bags. This second series of half-mil 'chutes

197

were made with diameters of 8 inches, 12 inches, 18 inches, 24 inches, 30 inches, 36 inches, 42 inches, and 48 inches.

A total of 240 drop tests were conducted by Kratzer with Greg Jones as timer and Bruce Blackistone as the loyal recovery crew. The 'chutes were hoisted back to the ceiling platform after each flight by a high-speed parachute crane consisting of a deep-sea fishing pole, line, and reel. Each of the twenty parachutes was tested with four different payload weights, giving a total of eighty different test combinations.

Basically, the tests confirmed experimentally what many model rocketeers already knew empirically. However, some interesting new data did result. For example:

1. The 8-inch 'chutes turned out to be drogue 'chutes. That is, they would act to stabilize the falling payload, but not to reduce its drop speed appreciably according to the drag equation. In some cases they did not open fully.

2. The performances of the hexagonal 'chutes and circular 'chutes were nearly identical.

3. Square 'chutes drifted least.

4. The second series (cleaner bag 'chutes) opened easier and more completely, were nearly approximate to the true hemispherical shape, and drifted more.

5. The square nondrifting 'chutes appeared to follow the drag equation very closely and exhibited a calculated C_d of 1.0 in neat accordance with theory.

6. However, as any competition modeler will tell you, the big half-mil cleaner bag 'chutes performed better. Their calculated C_d was as high as 2.25. Since this is an "absurdly high value" according to Malewicki, he goes on to state, "It tells us that the 'chute is gliding and generating lift in addition to drag."

To see how a parachute can generate lift, please refer to Figure 10-12. Remember that the big half-mil 'chutes drifted the most in the Kratzer drop tests. This sideways motion generated lift as shown in the diagram.

This leads at once to the questions: Can a model rocket parachute be designed and built in the same manner as a Para-Commander or other full-scale sky-diving sporting parachute? Can the performance of a parachute be improved by deliberately making it with unequal-length shroud lines to produce an eccentric load and therefore an induced angle of attack? Could the performance be improved by cutting a hole in one side of the canopy and venting the 'chute? If so, how big a hole and where and what shape? What is the effect of a vent cut in the precise center of the parachute? Does it stabilize the 'chute, and will it improve the performance?

The answers to all of the above questions are identical: We don't know yet.

It is very surprising that this information is not available in books and technical papers concerned with full-sized parachutes. One would certainly think that this research would have been done years ago! But it has not, and what little information is available is not readily applicable to model rocketry. Reasons: (1) model rocket parachutes are very much smaller than ordinary parachutes and therefore suffer greatly from scale effects, and (2) model parachutes are usually made from plastic film, which is nonporous, while all the available data relate to woven silk or nylon parachute canopies that let a certain amount of air bleed through them.

So there is a lot of basic parachute experimentation remaining to be done. The experiments would be simple, inexpensive, and easily conducted, but if properly designed and carefully carried out, they could give the answers to some important questions in model rocket recovery.

The basic rule of the NAR that determines parachute size required for safe, gentle recovery is: 10 square centimeters of parachute area per gram of recovered weight. This works out to 44 square inches of area per ounce of weight in the obsolete English system. Since the English system is more well-known to most of us at this time, you might remember that a 12-inch parachute is suitable for most models weighing up to 2 ounces, while a 24-inch parachute will handle most models up to 8 ounces.

Since no one has yet done any serious research and testing on the

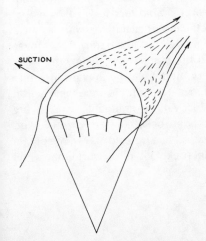

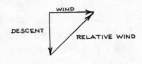

Figure 10-12: A parachute that moves sideways in the airstream will generate lift as shown.

effects of the shape of the recovered model dangling beneath the parachute and thereby affecting the airflow into the parachute canopy, we often run into some unusual situations. For example, I have had some designs that had a faster sink rate with a 24-inch 'chute than with an 18-inch 'chute.

Remembering the basic drag equation, you can easily understand that the drag of a 'chute—and therefore the descent rate of any given model—increases directly as the canopy area increases (and as the square of the linear dimension). Descent speed is a direct function of the drag of the parachute. The greater the 'chute area, the slower the descent rate. Doubling the linear dimensions decreases the descent rate by one-fourth!

This is an important factor to keep in mind because there are times when you do *not* want a slow descent. The launch site may be surrounded by rocket-eating trees, or it may be small. Or the wind may be blowing strongly. A large parachute will cause the model to drift for a long distance. On the other hand, when a model is carrying a fragile or heavy payload, a slow descent may be mandatory to prevent damage to the payload upon landing. Therefore, there can be no pat answer to the question of optimum 'chute size. It is subject to many trade-offs and compromises. This is why parachute duration competition is truly a sporting proposition that is *not* so simple and easy as it seems—especially when the contest rules require that the model be returned!

Other recovery devices

In 1959 Vernon D. Estes invented and perfected a model rocket that would fly in one direction only—up. *Tumble recovery* makes use of the motor ejection charge to kick the motor casing rearward where a hook catches and holds it in the model, but just barely. This moves the *CG* back behind the *CP*, and the model immediately becomes unstable. Thus, a tumble-recovery model will fly upward in stable flight, but will become unstable and start to tumble once the ejection charge activates. This type of recovery is limited to small, very lightweight models. Estes' first model rocket kit, the Astron Scout, was originally sold in 1961 and was still being sold at the time of this writing, making it the oldest model rocket kit in the world that is still available.

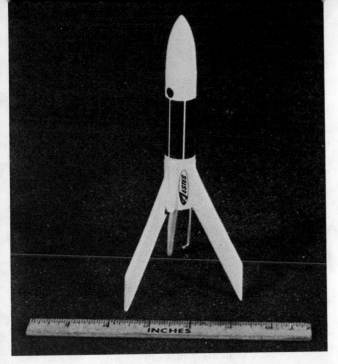

Figure 10-13: The Estes Astron Scout model rocket uses tumble recovery; the motor kicks back and brings the *CG* behind the *CP*. At this writing, this is the oldest continuously manufactured model rocket kit on the market, originating in 1961.

Featherweight recovery makes use of the drag equation, too. Here, the motor casing of a very small, very lightweight model rocket is ejected from the model. The model then falls very slowly because of its high area-to-weight ratio. It is literally like a feather. This recovery approach is also limited to very small and very lightweight model rockets. The NAR and the Federation Aeronautique Internationale (FAI) do not permit freely falling motor casings in competition, so featherweight recovery is not seen at contests. This system also suffers from the fact that once the motor ejects from the very small model high in the sky, the model becomes exceedingly difficult, if not impossible, to see. Most featherweight model rockets are one-flight jobs; you never see them again after ejection.

Recovery by autorotation is commonly thought of, but not so commonly used. The first such models ejected or released helicopterlike rotor blades that spun them slowly to the ground. Other models allowed their fins to cock to one side at motor ejection, thereby spinning the model down to the ground. These techniques are cute and interesting, but more difficult than the ubiquitous streamer and parachute recovery models.

And then there is *glide recovery.* . . .

Glide Recovery

From the very beginning of model rocketry in 1957 we early enthusiasts believed it would be possible to make a model rocket that would take off vertically, utilizing the high thrust of the model rocket motor, and return to the ground in gliding flight supported by wings or lifting surfaces. But we really didn't appreciate the problems involved in achieving this feat.

The early attempts were dismal failures, and the main reason was a lack of communication. In spite of the fact that model rocketry owes a great deal to model aeronautics, model rocketeers did not—and still don't—seem to come into the hobby by way of model aeronautics. For many years model rocketeers and model aviators didn't even talk to each other. The model rocketeers thought the model aviators were hopelessly outdated and old-fashioned with their propeller-powered flying machines. On the other hand, the model aviators thought of the model rocketeers as hopeless pyromaniacs who were creating huge explosions and clouds of smoke while cluttering up the skies with ugly plastic parachutes. These beliefs produced a lack of communication and forced model rocketeers to go through a long and painful period of trial-and-error experimentation, reinventing the wheel and making all the old mistakes over again. Model aviators could have explained a great deal about hand-launched and catapult-launched gliders to the model rocketeers.

When communications finally were established by some MIT model aviators who became interested in model rockets, the results were outstanding and spectacular. This should serve as a lesson to us all. Lack of communication, an inability or unwillingness to listen, and the failure to seek out all pertinent information can lead to a lack of progress in any field of human endeavor—not just model rocketry!

There have been rocket-powered model airplanes for a long time. Probably the first were flown in Bucharest, Rumania, in 1902 by Henri M. Coanda, who in 1962 watched me launch some modern glide-recovery model rockets! (History often has a strange thread that runs through it!)

Ron Moulton, editor of the outstanding British publication *Aeromodeller* and longtime supporter of "space modelling" in the Federation Aeronautique Internationale, attempted to build and fly glide-recovered aeromodels using skyrocket propulsion units in England in 1946. His efforts were frustrated by the lack of a small, reliable model rocket motor in those days.

Rocket-powered model airplanes became more common in about 1947 when the English Jetex motors were introduced. However, Jetex motors produce very low thrust and very long durations; they are also heavier than comparable model rocket motors. As a result, Jetex-powered models nearly always fly under power in a gentle, turning climb with their wings providing support against gravity. They are not normally capable of VTO (vertical takeoff).

Figure 11-1: A glide-recovery boost-glider lifts off at the Eleventh National Model Rocket Championships at the United States Air Force Academy.

Although model rocket motors of Type 1/4A through Type B do not have more total impulse than the Jetex motors, they do have much greater thrust and lower weight. Early attempts to substitute Type A5-3 model rocket motors for Jetex motors in model airplanes resulted in some spectacular failures. The models designed for Jetex power could not withstand the high accelerations and very high airspeeds produced by the model rocket motors. When the wings did not peel off, the resultant violent and rapid loop produced a rather hard prang, to use the British terminology for a flight that ends abruptly at a high angle to the ground.

The first publicly demonstrated glide-recovery model rocket was developed by Vernon D. Estes and John Schutz in 1961 and was flown at the Third National Model Rocket Championships in that year. The model was dubbed a boost-glider, or simply B/G for short, due to the two phases of its flight—VTO rocket boost followed by gliding recovery. The Estes-Schutz B/G is now in the National Air and Space Museum of the Smithsonian Institution, in Washington, D.C. It operated on a very simple principle. Basically, it was a gliding rocket. The wings were large fins at the aft end of the model, as shown in Figure 11-3. When the motor ejection charge went off, the expended motor casing was ejected. This triggered a mechanical latch that released control surfaces on the trailing edges of the wings. The removal of the motor casing weight plus the changing of the aerodynamic stabilizing surfaces produced a model rocket that would glide.

Once Estes and Schutz demonstrated a workable B/G model, many people soon proceeded to get a glide-recovery model into the air. This is another interesting phenomenon that often occurs in tech-

Figure 11-2: The late Dr. Henri M. Coanda (in black hat) flew the first rocket-powered glide models in Rumania in 1902. He visited a model rocket launching in 1962 where the author (left) and A. W. Guill (right) explained modern glide-recovery techniques.

Figure 11-3: The first modern boost-glider designed and built by Vernon D. Estes and John Schutz in 1961. Model is now in the Smithsonian Institution.

nology. A thing may be considered nearly impossible or very difficult until somebody finally gets something to work, crude as the device may be at the time. Once the breakthrough happens, developments follow very rapidly.

In the next five years the skies were filled with wild cut-and-try development with a number of different B/G configurations tested. Estes produced the Astron Space Plane B/G kit with control surfaces on the wings that flipped up when the expended motor casing was ejected from the model. Centuri Engineering Company soon followed with the delta-winged Aero-Bat, still an excellent B/G design. Paul Hans worked the bugs out of the canard configuration, and Ward Conley developed the first mini-birds. Hunt Evans Jõhnsen of MIT built the first swing-wing variable-geometry B/G, while Bill Barnitz of the USAF tackled the Rogallo flex-wing. All of us were reinventing the wheel, however, because we really had very little knowledge of why or how. It was empirical experimentation with little grounding in the theory of aerodynamics.

Then the second B/G breakthrough occurred. Larry Renger, then a senior at MIT majoring in aeronautical engineering, took a basic hand-launched glider and put a model rocket motor up front on a pylon. This was a conventional wings-and-tail airplane configuration. With the model rocket motor up front, the model was very nose-heavy and essentially was a ballistic vehicle. When the motor ejection charge went off, the expended motor casing was ejected. This removed weight from the nose of the glider and shifted the *CG* rearward. It also reduced the total weight of the glider. The glider then settled into a very nice high-efficiency glide. This was the famous Renger Sky Slash design. It was a rocket-powered glider, not a gliding model rocket. The emphasis was on boost-*glide*, rather than *boost*-glide as in previous B/Gs.

Properly trimmed, a Renger Sky Slash would outglide almost any other type of B/G then flying. It dominated B/G duration competition both here and abroad for years. Only recently has it been eclipsed by more sophisticated designs.

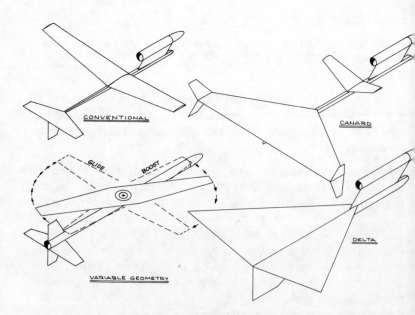

Figure 11-4: Basic glide-recovery model types.

Figure 11-5: The original Renger Sky Slash front-motored B/G as built from the Estes plans.

There is considerable question about who actually invented the pop pod for front-motored B/G models. The basic purpose of the pop pod is to remove as much weight as possible to permit better glide, and the obvious approach to doing this is to jettison everything that contributes to boost-phase and to retain only the glider portion for recovery. I was flying pop pods in April 1965, and I subsequently learned that Larry Renger had designed the FlaminGo, with pop pods, at about the same time. Here again is an example of parallel development in technology.

The next big breakthrough in B/Gs really wasn't a breakthrough. It could best be described as a smear-through. A B/G jettisons its

Figure 11-6: A boost-glider with a pop pod. The ejection charge of the motor kicks off the nose and ejects the pod streamer. Reaction kicks the pod backward and disengages a hook on the pod pylon.

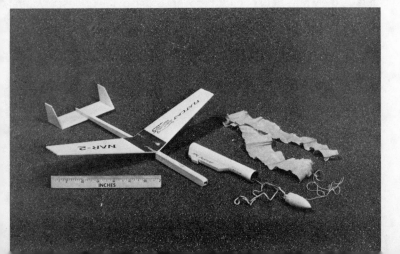

power package; or you might say that the rocket releases the glider it has carried aloft. Someone in the DelMarVa region around Washington, D.C., (I'll find out who no later than two days after this book is read by the person concerned!) came up with the concept of the *rocket glider*, or RG.

A rocket glider is a model rocket that is truly a glide-recovered model rocket. It cannot drop, or jettison, anything. What goes up must come down in one piece, together, in a glide, minus only the rocket propellant that took it aloft.

Now *this* was a real problem to design, build, and operate. There seem to be three ways to do the job:
1. Change the *CG* of the model by shifting the position of the motor casing at ejection or by designing to take into consideration the loss of weight due to the burned propellant.
2. Use the ejection charge to change the aerodynamic surfaces and thus change the trim of the model.
3. Use a combination of both.

There is considerable controversy, to which I am about to add, concerning whether or not the problems of RG have been satisfactorily solved as of this writing. An RG is an exceedingly difficult model to design, build, and fly. It is rare that two models of the same design built by the same modeler perform the same way or perform at all—to say nothing of two models built by separate modelers.

The RG fans will disagree violently, but, honestly, I have not seen the sort of reliable flight operation of RG models that I see regularly with B/G models. RG is a highly experimental area of model rocketry, one in which considerable thought, creativity, inspired design, and rigorous analysis remain to be carried out. It has not yet received the relentless scientific scrutiny given the B/G field.

In fact, the whole field of glide recovery is one of the hottest in model rocketry. Rapid progress is often made. Then years go by with no progress at all. It is also a specialized field with many little tricks, techniques, and procedures of its own. An entire book could now be devoted to glide recovery alone. Since we cannot possibly cover all aspects of this complex field, this chapter will have to serve primarily as an introduction to this fascinating phase of model rocketry.

Much of the following information is applicable to both boost-gliders and rocket gliders. Some is pertinent only to boost-gliders because we understand them better.

Flight of glide-recovery model rockets

Boost-gliders may be considered either as model rockets with gliding recovery or as gliders with rocket boost. Depending upon where you place the emphasis, they are *boost*-gliders or boost-*gliders*. The same holds true for rocket gliders.

The flight of any glide-recovery model rocket can be divided into two separate and distinct phases—the boost phase and the glide phase. The requirements for stability, structural integrity, and aerodynamics differ markedly from the boost phase to the glide phase. This simple fact means that a glide-recovery model must always be a compromise between a model rocket and a glider. The sort of technical trade-offs made by the designer dictates which type of glide-recovery model rocket the result will be.

The same simple fact also dictates that a change must take place for the model to make a successful transition between boost and glide phases. In a boost-glider the booster airframe and motor casing are jettisoned, causing a change in *CG* location, gliding weight of the model, or other factors. In a rocket glider the ejection charge of the motor must shift the position of the motor or part of the airframe or the setting of control surfaces.

Boost phase is that portion of the flight during which the model performs like a ballistic vehicle—a model rocket, if you will—launched from a standard launch pad with an electric ignition system. It is propelled aloft in a near-vertical trajectory by the high thrust of the model rocket motor. It trades vertical velocity and momentum for altitude in the time-honored fashion of a model rocket during the operation of the motor's time delay. During boost phase a glide-recovery model is not supported by lifting surfaces operating against the gravity vector force; whatever aerodynamic lifting surfaces are exposed to the airstream during boost must either have no effect upon the ballistic flight path of the model, or they must stabilize the model in the same manner as fins stabilize an ordinary model rocket.

The characteristics of the boost phase include both high accelerations and high airspeeds. These can combine to produce a situation aptly described by the model rocket term introduced by the Czechoslovaks—striptease. This is an understandable word in many languages! Large wings may rip off, and poorly made glue joints may turn loose. Therefore, glide-recovery models must be strong, and very good construction techniques must be used. The models usually have short, stubby wings in comparison to nonboosted gliders and sailplanes.

The glide phase has entirely different characteristics. The model in a glide is supported by the lifting force of its wings, which sustain it against gravity by the forward airspeed creating airflow over the wings. This airspeed is as low as 10 feet per second or less. The sink rate, or vertical descent rate, may be as low as 1 foot per second. Since the sink rate of a glider is directly related to the ratio between the lift and drag of the glider, very high lift-to-drag ratios on the order of 10 or more are desirable—if you can obtain them!

As during boost phase, the model must be stable in pitch, yaw, and roll, but the pitch and yaw axes can no longer be considered identical because most glide-recovery models are not symmetrical in the pitch and yaw axes. The glide-recovery model must have very high inherent stability because winds, gusts, and thermals (rising air columns) will continually disturb it during glide. If it pitches nose-up, the forces generated by its lifting-stabilizing surfaces must bring the nose down into an equilibrium gliding condition again. If the nose pitches down, forces must be generated to bring the nose up again and terminate the dive. If the model rolls, it must right itself—and it must "know" what is right-side-up. If it yaws left or right, the yaw must be corrected by stabilizing surfaces.

A fin-stabilized model rocket is not a good glider, and a glider is not a good model rocket. Since the two flight phases are very different in their requirements for design and construction, compromises must be made. If maximum flight duration is the objective, as it is in competition, it can be obtained in either of two ways. A modeler can strive for maximum altitude during boost, compromising the glide characteristics and obtaining long flight duration by simply having the model take a very long time to descend from a high altitude in spite of a very high sink rate. Or the model rocket designer can work toward maximum performance during the glide phase, cutting back the altitude gained during boost and striving for a very low sink rate.

A great deal depends upon the weather at the launch site when you get ready to fly. Many modelers take two different glide-recovery models to a contest—a foul-weather model and a fair-weather model. A foul-weather model usually has good boost and very stable glide, while a fair-weather B/G usually is designed for absolute maximum glide performance.

Stability and control

Except for the very few radio-controlled glide-recovery models that exist at the time of this writing, we cannot put a pilot in our models to keep them under control during their flight. Therefore, glide-recovery models must have inherently good stability and be able to maintain their equilibrium during flight because of this good stability.

Roll stability during boost phase may not be required. In fact, some modelers deliberately induce a low roll rate during boost by means of slightly offset pods or motor thrust lines. This causes the model to "screw" its way up into the air. Basically, any pitch or yaw motions are averaged out by rolling the model during boost phase. Too much boost-phase roll can be disastrous, however, because it so reduces the altitude that a long glide cannot be obtained. During glide phase roll stability can be obtained by using a dihedral angle on the wings,

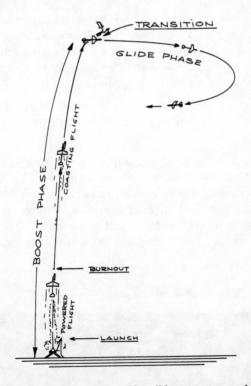

Figure 11-7: The basic flight phases of a glide-recovery model.

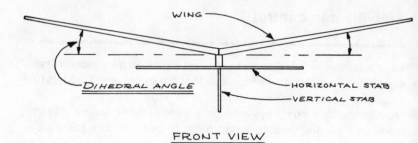

FRONT VIEW

Figure 11-8: Dihedral angle seen from front.

as shown in Figure 11-8. This dihedral angle should be between 15 degrees and 20 degrees per wing. It can be made up of compound dihedral angles—a straight center section with the tips turned up to the same location they would have if the wing had a straight dihedral angle. The optimum dihedral for a wing is actually elliptical, as shown in Figure 11-9.

Stability in the yaw axis is obtained by rudders or vertical stabilizing surfaces. Normally adequate yaw stability can be obtained by having a vertical stab area equal to 5% to 10% of the wing area. However, the location of the vertical stab often has a lot to do with both yaw and pitch stability, as we will see later.

The biggest control and stability problem of all glide-recovery models in boost phase and glide phase is in the pitch axis. See Figure 11-10.

During boost phase an improperly designed, constructed, or balanced glide-recovery model—except parasite types that we will discuss later—will pitch up into a loop or pitch down into an outside loop. Although these aerobatics are often spectacular, the model usually prangs with great force and is demolished. Or it loops so hard and tight that it does not gain sufficient altitude for a long glide.

Figure 11-9: Elliptical dihedral of a wing.

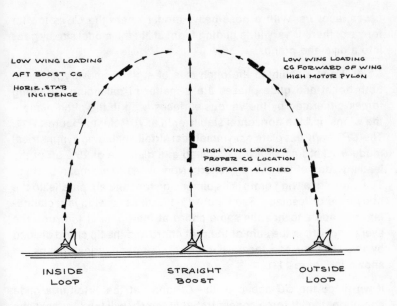

LOW WING LOADING
AFT BOOST CG
HORIZ. STAB
INCIDENCE

LOW WING LOADING
CG FORWARD OF WING
HIGH MOTOR PYLON

HIGH WING LOADING
PROPER CG LOCATION
SURFACES ALIGNED

INSIDE
LOOP

STRAIGHT
BOOST

OUTSIDE
LOOP

Figure 11-10: Pitch axis stability problems of glide-recovery model rockets in boost phase.

The radius of the loop caused by pitch depends upon the location of the *CG* and the wing loading (ounces of weight per square inch of wing area). The farther back the *CG* or the lower the wing loading, the smaller the loop radius. Therefore, the most desirable characteristics during boost phase are forward location of the *CG* and high wing loading—that is, more weight per unit wing area.

This is the reason a parasite glider—a gliding portion of a boost-glider that is hung on the side of a much larger booster vehicle and therefore has a very high wing loading while attached to the booster—works so well during boost. It is also the reason a front-motored B/G model performs better than a rear-motored design.

Improper balance, or trim (location of the *CG*), during glide phase will cause a glide-recovery model to pitch up into a stall or pitch down and dive into the ground. A badly trimmed model may also go into increasingly more violent stalls, ending up in a spiral dive into the ground.

Stalls are caused by the *CG* being too far aft—a tail-heavy model, in other words. At its worst this aft *CG* condition leads to sharper and deeper stalls until the model goes into a spiral dive from which it never recovers. With a nose-heavy model where the *CG* is too far forward, there is very little gliding flight at all; the model simply goes into a dive and prangs.

Control and stability in the pitch axis of a glide-recovery model in both boost and glide phases are a matter of balancing the lifting forces generated by the various surfaces lying in the pitch plane—the wings and the horizontal stabilizer (if any). Refer to Figure 11-11. The *CP* of *any* flat plate or symmetrical airfoil, including symmetrical model rocket fins, is always located at a distance of 25% aft of the leading edge of the average chord. Now, what do we mean by that? The chord of a wing or fin is its dimension fore and aft, parallel to the body tube or fuselage. See Figure 11-12. Unless a wing is rectangular in shape and has the same chord at the tip as at the root, the average chord is the sum of the root chord and the tip chord divided by two. Or you can locate it precisely by geometrical means, as shown in Figure 11-13.

If we place the *CG* ahead of the *CP* and put the airfoil in a glide condition, the lift force concentrated at the *CP* will tend to make the wing pitch down, as shown in Figure 11-11. There is no lift force to make it pitch up again. So it simply dives into the ground. This is precisely the condition wanted for a good fin-stabilized model rocket, but not for a glide-recovery model rocket during glide phase!

Therefore, we must have some sort of mechanism that will bring the nose up when the model starts to dive and, conversely, bring the

Figure 11-11: Cross section of a symmetrical airfoil with *CP* located at 25% of chord and *CG* located ahead of *CP*. Pitching moment will always be in the direction of the oncoming airstream, making the airfoil always point into the airstream. This is a stable condition for boost phase but will not result in a glide that is stable.

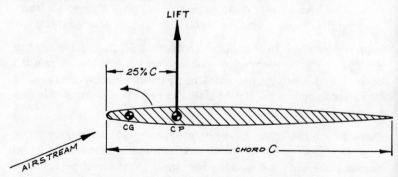

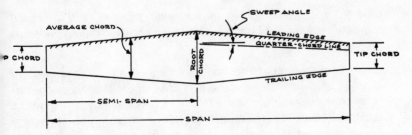

Figure 11-12: Definitions of the dimensions of a wing.

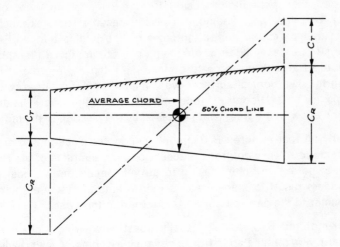

Figure 11-13: How to locate the average chord by geometric means.

nose down when the model starts to climb and enter a stall. We must obtain a *balance of forces*. Our glide-recovery model is just like a teeter-totter with the balance point at the model's *CG*. Anything that makes it want to dive should immediately produce a force to bring it back to level flight again. Anything that makes it want to climb should produce a force that will bring the nose down to level flight. In other words, it must be *stable* in the pitch axis just as it is stable in the roll axis—thanks to wing dihedral—and in the yaw axis—thanks to the vertical stab. Any tendency of the model to wander away from its balanced, stable condition must be counteracted by forces that will return it to a balanced glide condition.

As you can see, this stable glide condition is radically different from the stability condition required during boost phase. Therefore, something must happen to the model to convert it from a ballistic, fin-stabilized model rocket into a gliding model, and it should logically occur at or near the peak altitude achieved during the boost phase.

Glide-recovery model rockets are designed so that the ejection charges of their motors cause a change in the models to convert them from rockets to gliders. This is usually done by changing the location of the CG of the model, or changing the physical configuration of the model, or both.

Changing the physical configuration of the model was the method first used by Vernon D. Estes and John Schutz. The ejection charge ejected the spent motor casing from the model; this not only decreased the wing loading, but also released a complex mechanism that permitted control surfaces on the trailing edges of the wings to be deflected upward. As you can see in Figure 11-14, this produced a pitch-up force to oppose the pitch-down force of the wing lift. The control surfaces were hinged to the wings by cloth strips as in a U-control model airplane, and they were held in alignment with the wings until released by the ejection of the motor.

Nearly all rear-motored B/G models use motor ejection or similar systems to release control surfaces. However, the rules of the NAR do not permit an empty casing to be jettisoned from a model, so modelers resort to many clever arrangements to attach a colored streamer to the casing in order to conform to the rules.

Rear-motored B/G models usually boast flying-wing, delta-wing, canard (horizontal stab in front), or variable-geometry (swing-wing) configurations. A number of highly successful front-motored and rear-motored swing-wings have been built and flown, as well as successful models that change configuration. However, the simplest of all B/G models to design and build—but not to trim and fly—is the front-motored, pop-pod, conventional configuration rocket-boosted glider, which is shown in Figure 11-15.

With movable-surface glide-recovery models, we saw how the lift force generated by the wing was counterbalanced by the lift force generated by control surfaces to maintain a stable glide. In these cases the CG was well forward, usually located at less than 25% of the average wing chord.

Front-motored, pop-pod B/G models and front-motored, motor-shifting RG models usually do not change any aerodynamic surfaces when the motor ejection charge goes off. Instead, the CG of the model is shifted, and in the case of B/G models the wing loading is

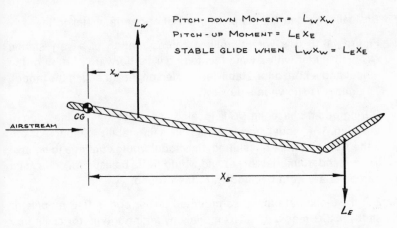

Figure 11-14: Use of an elevon control surface on a flying wing to provide stability. Wing alone gives pitch-down moment; elevon gives pitch-up moment. When the two pitching moments are equal, the wing will glide in a stable condition.

Figure 11-15: A typical front-motored conventional configuration boost-glider, the author's contest-winning Unicorn design with Vee-tail and pop pod.

greatly reduced by the jettisoning of the power pod—or the separation of the booster rocket in the case of parasite B/G models. Thus, in *CG*-shift models no new lifting forces are created at the transition from boost to glide, but the *relationships* between the lifting forces and the *CG* are changed by changing the location of the *CG*.

Basically, we change the pivot point of our aerial teeter-totter.

Figure 11-16 is a very simplified representation of a very simple *CG*-shift glider with a wing located near the forward portion of the model and a horizontal stabilizer located on the aft end of the model. The glider is shown in side view.

During boost phase the *CG* is located at or near the leading edge of the wing. The model therefore acts like a fin-stabilized model rocket. If the *CG* point were not shifted, the model would continue to behave like a standard model rocket and would fly its ballistic trajectory in a parabolic arc right back down to the ground.

But if we separate the nose-mounted power pod in B/G models or shift the *CG* rearward in RG models by letting the motor casing be forced to the rear and caught there, we have a different situation entirely. Let us suppose that the *CG* is shifted aft to a location at 50% of the wing chord. The *CG* is now *behind* the *CP* of the wing and *ahead* of the *CP* of the horizontal stab.

If the wing were there all alone, the wing lift concentrated at the *CP* would cause the wing to pitch up. You can create this situation by taking a piece of 1/16-inch sheet balsa 3 inches wide by 12 inches long; try to make it glide by itself. You will find that its leading edge will pitch up, and the whole sheet will descend to the floor in a rotational flip, going end-over-end and rotating rapidly about its *CG* point at 50% of the chord.

Figure 11-16: Representation of a *CG*-shift boost-glider in gliding condition with *CG* shifted aft to 50% of the chord. The wing now has pitch-up moment that is balanced by the pitch-down moment generated by the aft-mounted horizontal stabilizer.

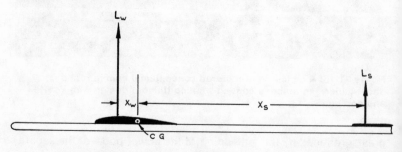

$$\text{PITCH-UP MOMENT} = L_w X_w$$
$$\text{PITCH-DOWN MOMENT} = L_s X_s$$
$$\text{STABLE GLIDE WHEN } L_w X_w = L_s X_s$$

Therefore, a counteracting pitch-down force must be created to keep the model from pitching up. This is done by adding the horizontal stab to the rear end of a long fuselage. The horizontal stab is really just a little wing. It also produces lift, but the *CG* is a long distance ahead of its *CP*. This lifting force of the horizontal stab produces the pitch-down force.

Although the lift force of the wing is much greater than that of the stab because of the greater area of the wing, the lifting force of the wing is closer to the *CG*. This is like balancing a 50-pound boy and a 250-pound middle linebacker on a teeter-totter; the 250-pound man is going to be closer to the pivot than the 50-pound boy. But their *moments* will be equal.

What is a *moment*? Simply the product of the force and the distance through which it acts. It is the weight times the distance from the balance point—whether it be on a teeter-totter or in a gliding B/G. As you can see from Figure 11-16, the moment of the stab (a small force acting over a long distance) is equal to the moment of the wing (a large force acting over a short distance).

If the model starts to dive, the wing builds up far more lift than the stab and therefore causes the model to pitch up again into its trimmed glide angle. If the model starts to climb, the wing loses lift, and the stab therefore causes the model to pitch down into a stable glide.

Now, it is possible to have the wing and the stab exactly the same size. This is called a tandem wing model. But the glide *CG* point would have to be about halfway between the two wings.

A canard model with a small horizontal stab in front and a big wing in the rear can also be thought of as a conventional model with a very tiny wing and a very big stab. The *CG* must be even farther aft for a canard model.

This is a very simplified explanation of glide structures and dynamics. I had to digest a lot of data from NASA reports and books before boiling it down to this point. But you do not have to be an aerodynamicist with a degree to design a good glide-recovery model, nor do you have to be an expert to build and fly one. But you do have to be careful in your workmanship, because this is a very complex area of model rocketry, as you have probably surmised by this time!

To design a *CG*-shift glide-recovery model from scratch, one should have access to airfoil data such as in *The Theory of Wing Sections* by

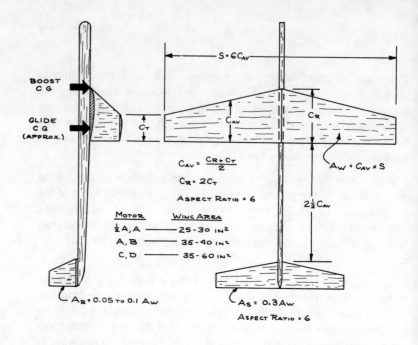

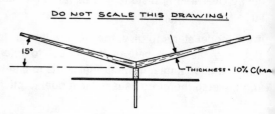

Figure 11-17: Stine's Basic B/G Design Rules.

Abbott and von Doenhoff (see Bibliography), which is a standard manual for aeronautical engineers. The method used to make the calculations can be found in Frank Zaic's excellent and informative book *Circular Airflow and Model Aircraft*, also listed in the Bibliography. However, you do not need to delve into these sources unless you want to, because I have worked out the following set of empirical design rules of thumb for front-motored *CG*-shift B/G or RG models. The rules are not arbitrary. They were developed over a period of years from a careful study of NASA wind tunnel data, and they are soundly based in aerodynamic theory. They have been put to the test by modelers young and old, tyro and advanced. They work. If you do not want to try a kit but prefer to go it alone, they will work for you if you follow them carefully.

The basic Stine Design Rules are shown in Figure 11-17. This is a simplified three-view drawing of a hypothetical front-motored conventional configuration.

You start with the basic dimension—average wing chord. The wing span should be five to eight times this number, meaning an aspect ratio of from five to eight for those readers who are also model airplane builders; if you are not, the meaning of aspect ratio is shown in Figure 11-18.

For optimum glide performance a wing with very little sweepback—less than 20 degrees—should be used, even though it doesn't look

Figure 11-18: Aspect ratio of a wing or fin.

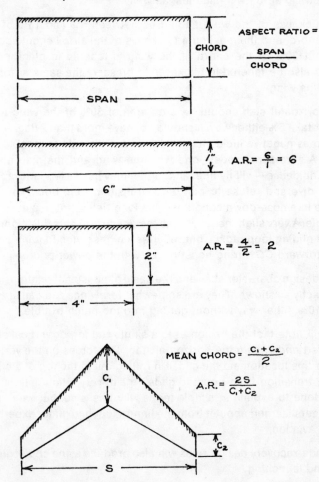

very fast and sleek. High sweepback angles create high induced drag. Remember that a high-performance sailplane has long, thin wings. Although the elliptical wing planform such as that used on the legendary Supermarine Spitfire is almost the optimum aerodynamic wing shape, this optimum shape can be approximated within 2% by using a straight taper wing, as shown. The taper wing is similar to that used on an equally legendary airplane, the North American P-51 Mustang fighter plane of the same era.

Although a glide-recovery model will fly with an unstreamlined slab of sheet balsa for a wing, it will fly better and be much easier to trim for glide if an airfoil is used, as shown. The classical airfoil that is curved on top and flat on the bottom isn't really necessary; a purely symmetrical airfoil will glide just as well.

The thickness of the wing should be between 5% and 15% of the wing chord, and the maximum thickness of the airfoil should occur at about 30% of the chord. If the wing is tapered in planform, it should also be tapered in thickness to preserve the same airfoil all along the wing.

The horizontal stab should have an area of 30% of the wing area, give or take 5% either way. It should not have more than 2 degrees to 3 degrees negative incidence, or downward tilt, with respect to the wing. A zero-zero model—one with the wing and the stab at the same incidence—will fly fine, although it may have a tendency to get into a dive and refuse to pull out if transition from boost to glide occurs in a nose-down condition—if a long-delay motor is used, for example. A very slight negative stab incidence prevents this, tightens up the gliding loop radius, but requires a higher wing loading and a more forward *CG* during boost phase with the power pod on.

Other design characteristics and the basic dimensional relationships are exactly as shown. They are somewhat forgiving. You can stretch them 10% either way without getting into too much trouble.

You will note that the design uses a single rudder, or vertical stab, mounted *underneath* the horizontal stab. The forces on the vertical stab in this location aid the glider in rolling out at the top of a climb during transition from boost to glide. Again, very little analysis has been done to explain in simple terms why this is so, but we model rocketeers learned about it from the hand-launched glider experts in model aviation.

The glide-recovery design rules will also produce a fine chuck glider for hand launching.

Figure 11-19: Nose-heavy gliding flight.

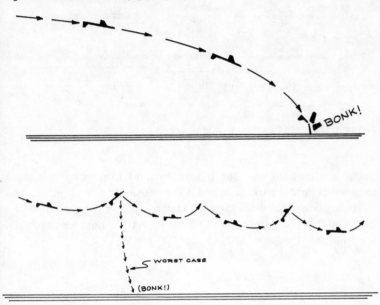

Figure 11-20: Stalling, tail-heavy gliding flight.

When completed, the basic glider should have a weight of about 0.025 ounce (0.71 gram) per square inch of wing area for a good glide.

Finishing a glide-recovery model is another area of great controversy in which there is yet no concrete data to point one way or the other. Some modelers leave their gliders unpainted with the glider sanded smooth using No. 400 sandpaper. Others use glider dope, a mixture of 50% clear dope and 50% dope thinner, putting on several coats and sanding with No. 400 sandpaper between coats until a smooth finish is obtained. Still others use filler to remove all balsa grain and then paint the model with several coats of dope or acrylic enamel with a spray gun or airbrush.

Although rocket glider models retain their motor casings during glide, nearly all competition B/Gs in the lower classes powered by Type C motors or smaller use what is known as the pop pod. A typical pop pod is shown in Figure 11-21. A pop pod separates the entire weight of the propulsion unit from the glider at transition. The unit shown incorporates a streamer for recovery. Some of the larger

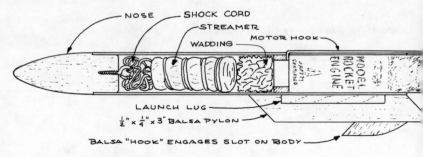

Figure 11-21: Cross section of a typical pop pod.

pods use parachutes. But I have learned from long and hard experience that a parachute will often cause the pod to get tangled with the glider. You want that pod to drop out of the glide path of the glider so that the glider doesn't run into it. A parachute tends to keep it up there in the glide path.

Trimming

A glide-recovery model must be trimmed for glide before it is flown. There are very few glide-recovery models that can be flown without glide trimming. The exceptions to this rule are the delta-wing types.

Trimming a B/G is easy. Use the glider only without the pop pod mounted, or without the motor casing in it if you do not use a pop pod. In other words, the glider portion must be trimmed in its gliding configuration. A grass-covered field is best for trimming because it keeps the glider from getting dinged if it dives into the ground. Naturally, if there is one rock in the field, the glider will be drawn to it like a magnet and hit it. And, naturally, the glider will always break in the worst possible place!

The first trim glides should be gentle. Grasp the glider behind the wings or in a place near the *CG* where you can get a good grip. Toss it *gently* with an overhand motion away from you, turning it loose into a glide path slightly nose down. Don't heave it! Make several tosses; you may not have tossed it correctly the first time.

If the glider dives, as shown in Figure 11-19, it is nose-heavy. You must therefore remove weight from the nose or add some to the tail, or both. If the glider stalls, as shown in Figure 11-20, it is tail-heavy, and you will have to add weight to the nose.

The best thing to use for glide trim weight is a little glob of plasticene modeling clay that can usually be purchased in the crafts section of a hobby store or toy store. Put some in your range kit. To use it, pinch a little onto the nose or tail of the glider, as the case warrants. Some glider designs have trim weight compartments built into them so that you do not end up with a huge wad of clay on the nose, which may create as much drag as the rest of the glider. Add a little clay at a time until, finally, the glider sails away from you in a slowly descending glide path, landing about 30 feet or more away.

Now add just a teeny pinch of clay to the *left* wing tip to cause the glider to turn slowly to the left as it glides. Reason for the turn: Once a glider gets a couple of hundred feet up after its boost phase, you do not want it to sail off on a straight cross-country flight. A straight-flying glider will nearly always turn its tail into the wind and take off for the next state, going downwind much faster than you can possibly run. What you want is a glider that turns so it will circle over the launch area.

Why turn left? On warm and hot days rising bubbles of hot air called thermals are generated in most open fields. If your glider gets caught in one of these thermals, you may well lose it. A thermal is a miniature low-pressure cell like a miniature storm system, except it doesn't have clouds in it. It is a doughnut of air that circles clockwise when viewed from below. A glider that turns to the right will get into a thermal and turn with it, riding it up and up. A left-turning glider will fly out of a thermal sooner, and you have the chance to get it back again.

Once having trimmed in a gentle glide, you are ready for a heave test. Haul off and heave the glider straight up as hard as you can. The wings stayed on? Good! The glider should climb straight up and eventually lose speed, roll over into a glide, and glide. If not, try again. If the model won't perform like this, remove some nose weight and make it slightly tail-heavy—but just slightly. Some gliders will not pass the heave test no matter how hard you try. So, if after persistent attempts to get a successful heave test, you cannot get the blasted thing to roll out and glide, go ahead and try it under power.

Now balance the model for boost flight. Add the pop pod and/or motor. The model should balance at the leading edge of the wing. If it does not, add or remove weight from the *pop pod* on the nose of the B/G. If it is a surface-change B/G, check that the glide control surfaces are locked in the proper position for flight. If in doubt with a kit model, read the instructions!

You are now ready to launch. Make sure that the pop pod is free enough to come loose at transition when the action of the ejection charge kicks it off. In other types of B/G models make sure that the transition mechanism, if any, is free to work as it is supposed to.

A B/G normally needs to be held up off the launch pad base or deflector. A clothespin or a piece of tape wrapped around the launch rod will support the model properly.

Use an umbilical mast to hold the electrical leads for front-motored B/G models. This is shown in Figure 11-22. An umbilical mast allows the leads to fall free *away* from the model so that they don't get caught in the model's tail section. You can also tape the electrical leads to the launch rod to keep them from falling across the horizontal stab. It is very embarrassing to have this happen, especially in a contest where tangled leads count for an official flight.

It is always a good idea to have somebody to help you on recovery. Your partner should go after the pop pod or motor section while you chase the glider. Don't count on other range people to watch your pod or booster; they see far too many models fly and land, and they quickly forget where.

Figure 11-22: An umbilical mast installed on a launcher to support ignition wires and prevent them from fouling the tail section of a B/G at lift-off.

Figure 11-23: One of the most unusual B/G models ever flown was Douglas J. Malewicki's legendary Snoopy, shown here with Mike Poss and a genuine Snoopy dog house launcher at NARAM-9. Snoopy's ears were wings. Hind paws became horizontal stab.

Trimming an RG model is similar to trimming a B/G. Trim for glide first. You must do this with an empty motor casing in the location it is supposed to maintain during glide. Sometimes it holds the proper position, and sometimes it doesn't. Because of its higher wing loading, today's RG model glides faster than an equivalent B/G, so you may not be able to glide test it in any other way than throwing it out of a second-story window.

Carefully check the boost-phase balance of any RG. This is where most RG models get into big trouble. Therefore, the first flight of an RG model should be a heads-up affair; RGs have been known to loop violently and attempt to part the hair of the modeler or bystanders. RG modeling is still a highly experimental area of model rocketry and should be engaged in with more than reasonable care.

Radio-controlled glide-recovery models

One of the first questions a person asks about model rockets is: How about installing radio control? It's possible, but so far practical only in glide-recovery models.

During powered, or boost, flight things happen far too fast for a pilot on the ground to exercise effective control over a normal model rocket. With a glide-recovery model rocket, however, one may certainly control the glide phase and, to a lesser extent, the boost phase as well. The reason is that glide-recovery models usually fly at lower airspeeds than ordinary model rockets.

Single-channel, rudder-only radio control has been flown with great success in B/G work. At the Twelfth National Model Rocket Championships at the Manned Spacecraft Center in Houston, Texas, Doug Malewicki sprung a radio-controlled B/G on everyone for the spot landing event. He steered his model right down to the target, much to my own consternation because my model had been closest to the target pole until that time. Such micro-miniature radio-control work requires the use of extremely small receivers and servos. But it can be done, and it is a real challenge for the advanced modeler.

Multi-channel radio control was featured in Larry Renger's Sky Dancer B/G, which was powered by a Type D motor and was also flown at the Twelfth National meet. This model had a "full house" installed and was completely controllable during both boost and glide—and it was a very large model! Details were published in *American Aircraft Modeler* magazine in 1970. Renger's model was a very expensive project with a price tag of several hundred dollars for the radio-control equipment alone.

Recap

Glide-recovery model rockets make up a fast-moving, highly advanced segment of model rocketry. They are tremendously challenging. Anyone who thinks there is nothing to model rocketry but "up and down" has not tangled with glide recovery. It is a very controversial field. Some parts of it are well-researched; other parts operate on a highly experimental, cut-and-try basis. Developments take place rapidly. It is *not* a beginner's field of endeavor. A modeler must have great skill, patience, craftsmanship, and creativity. There is much left to be done, much yet to be learned, and much remaining to be researched and tested. All of the answers are not yet in by any means.

And glide recovery is a lot of fun! Extremely popular in Europe, it is spreading ever more rapidly in the United States.

Building and Flying Large Models

Some model rocketeers enjoy building and flying large model rockets powered by Type D, Type E, and Type F motors. These larger model rockets take longer to build and are much more expensive than the smaller models that have been discussed. They require more clear, open land area for flying. Because of the increased amount of rocket propellant in the larger motors, the higher weights of the models, the higher velocities attained, and the greater altitudes possible, a constant and conscientious observance of the strictest safety rules must be followed at all times.

I have emphasized safety codes so strongly throughout this book because I want them to become second nature to you. Even when flying the smallest models, you should be following the safety rules automatically because they are good habits that you have formed. And if you fly the larger models, you will need every bit of safety consciousness and safety training that you have! Some large model rockets are as big as small professional weather rockets and research rockets, and birds that size can be very tricky!

Everything that applies to other model rockets also applies to large model rockets—and often more so. You may wish to go back and review some of the earlier chapters as you get into this one. Go ahead. There is no need to be embarrassed about doing it. Better to review than to have a disaster with a twenty-dollar model rocket traveling at the speed of sound.

Construction

Most large model rockets use the same construction techniques and materials as the smaller ones. However, special care must be taken to make the model strong and durable. A large model propelled by a model rocket motor of greater than 20 newton-seconds total impulse will undergo accelerations ranging from 10 g's to 50 g's during powered flight. It will achieve very high airspeeds, perhaps approaching Mach-1 (the speed of sound). Therefore, the aerodynamic loads on it will be very high because, if you remember, the drag forces increase as the square of the velocity. Some of these large models will go four to five times faster than their smaller brothers; this means that the forces on their fins are sixteen to twenty-five times as great!

So, large models must be built *strongly*. Make good glue joints. Use proper materials. Do a careful job.

Although most large models have body tubes in the 1-inch to 2-inch diameter range, sturdy spiral-wound paper body tubes up to nearly 4 inches in diameter are now available from model rocket manufacturers. Strong yet lightweight, the spiral-wound body tubes can be further strengthened by wrapping them with wet, doped model airplane tissue or silkspan. Or they can be covered with a heat-shrunk layer of Monokote or other model airplane plastic covering material.

Large noses fabricated of plastic or balsa are also on the market. However, when one begins to build large models, a lathe is often a very handy tool to have available.

With a lathe, you can turn your own noses and boattails from balsa blocks. It is also possible to turn complete model rocket bodies from block balsa, using the hollow log method of construction. Turn the body from block balsa on a lathe. Cut it in half lengthwise. Hollow out both halves. Locate a paper body tube down the center. Glue the halves back together again. Sand and fill the joint, and you will never be able to tell it is there! This technique has been brought to a high pitch of perfection by Don Sahlin, master puppeteer who formerly constructed all the puppets and marionettes seen in Jim Henson's "Muppets" and "Sesame Street." A longtime model rocketeer, Sahlin used his knowledge of lightweight wooden construction techniques to build some beautiful large models. One is shown in photographs later in this chapter.

Figure 12-1: Large model rockets are capable of very high performances and can carry large payloads. They should be flown only by advanced model rocketeers.

In the very large-diameter bodies it is most important to use a stuffer tube to restrict the internal volume of the body tube that must be pressurized by the motor ejection charge. Often, the body tube is so large that the puff of gas from the ejection charge can't build up enough pressure to pop the nose and eject the recovery 'chute. Many excellent large models have pranged because the ejection charge failed to pop the nose. A stuffer tube is a smaller tube, usually 3/4 inch to 1 inch in diameter, that leads from the motor thrust mount forward to the base of a parachute compartment, which is a larger internal tube that houses the 'chute. This stuffer tube ducts all of the ejection charge gases into the parachute compartment. The large, empty part of the body tube is not pressurized. A typical stuffer tube installation is shown in Figure 12-2.

The motor mount must also seal off the ejection charge to prevent it from escaping to the rear around the motor casing, thus failing to pressurize the body tube and parachute compartment. Motor mounts for these very large motors and models are available for most body tube sizes. But you may have to make your own. If you build a motor mount for any tube larger than 2 inches in diameter, I recommend that you use model airplane 1/16-inch plywood rather than cardboard or paper; you will need the extra-strong mount to withstand the thrust force of the large motor.

Large models often have heavy noses full of payload or ballast weight. When the noses are popped off in flight and come up against the end of the shock cord, there is a considerable jerk. So use a heavy *cotton* line, and don't be afraid to make it 3 to 4 feet long—or longer. The extra length will give the nose time to slow down before it takes up all the slack in the shock cord. Shock cord attachments must be strong to withstand this sort of jerk. A simple screw eye in

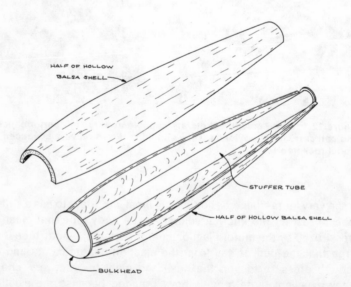

Figure 12-2: The hollow log technique of building model bodies. The exterior shape is first turned on a lathe; then the body is cut in half lengthwise. The halves are hollowed out, and a body tube, or stuffer tube, is placed down the center. The halves are then glued back together and the joint sanded and filled.

the base of a balsa nose will usually pull out. So inset a piece of 1/4-inch or 1/2-inch wooden dowel into the base of the nose and screw the screw eye into that.

Although I have built large models with 1/16-inch sheet balsa fins, I do not recommend this practice unless you really know what you are doing in the workshop. As a rule, use at least 3/32-inch sheet balsa for fins. Better yet, use 1/8-inch sheet balsa. If you put a good airfoil on these thick fins, as discussed in earlier chapters, you will not increase the total drag more than a couple of percentage points, and you will gain by having a greater area at the fin root to glue to the body tube. You will also increase the strength of the fins, of course. If you prefer flying with very thin airfoil sections, use model airplane plywood for the fins.

Because body-fin joints on large models should be very strong, it is mandatory that double-glue joints be made—and made well. It is also wise to reinforce the joints with model airplane tissue or silkspan, as shown in Figure 12-3. For maximum strength run the silkspan completely over the fin surface and around the body tube. This will not add very much weight, and it will increase the strength enormously.

To prevent launch lugs from breaking off, you can also dope a piece of silkspan over them, as shown in Figure 12-3.

Because an unstable model rocket with a Type F motor is downright

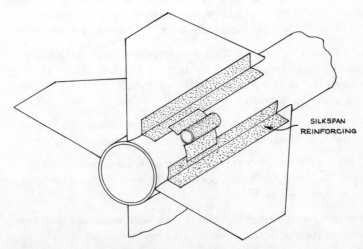

Figure 12-3: Reinforcing fin and lug joints with silkspan or model airplane tissue.

terrifying and even potentially deadly, check and recheck your *CP-CG* calculations. Give the model the swing test. Make sure it is stable.

It is especially important to build a strong and very lightweight model if you are going to fly it with one of the low-thrust, long-duration motors such as the Flight Systems, Inc., Type F7. The low thrust (about 1 pound) and the very long duration (about 9 seconds) of this motor, which has been nicknamed "The Steam Machine" by respectful advanced model rocketeers, produce a very slow lift-off and climb. A heavy model weighing more than 8 ounces is likely to pitch over under thrust into a gravity turn, which will cause it to arc into the ground under power.

Whatever you usually do in making small models, do it double for the big ones. Make sure that all glue joints are good. Make sure that the *CP* and *CG* are correctly placed. Make sure that the recovery device, chute or streamer, is strong and installed well. Do not take shortcuts or do sloppy work on big models. They can bite.

Flight operations

Although I have launched Type F models from a standard 1/8-inch launch rod 36 inches long, I prefer to launch them from a 3/16-inch diameter rod 60 inches long. Rail launchers are also recommended for large models because rods can whip under the heavy weight.

Use a good electrical system with large models. It is wise to have at least 20 feet of wire between you and the model. If something goes wrong, that is a very large model rocket motor out there on the pad, and distance is safety.

Follow the ignition instructions of the motor manufacturer. Some large motors require special igniters and ignition techniques.

Carry out your preflight prepping with extra-special care. Have somebody double-check your work. There is no room for error with large model rockets.

You will need a large flying area. The school football field isn't big enough for any motor larger than a Type D. Get out of town and into the country. I would not recommend launching large models from fields that are smaller than 1,000 feet on a side. This is a little more than three football fields end-to-end.

Tilt your launch rod or rail about 5 degrees away from the vertical and toward an open and unpeopled downrange area where the model can prang if something goes wrong. The downrange recovery area should be completely clear of people, and you should keep it clear during flight operations! Don't let a recovery crew wander about out there.

Although smaller model rockets can be placed on launch pads 2 to 3 feet apart with safety, try to have at least 10 feet between launch pads for large models. Sometimes the jet backwash from the lift-off of large models is substantial.

The area around a launch pad with a large model on it should be cleared of people for a distance of 25 feet. Even launch personnel should not be near the model. All people around the launch area should be aware of what is going on and should be in an alert heads-up condition during the final countdown. Nobody should be lying on the ground. Motor storage boxes should be closed. Everyone's attention should be on the big bird.

The lift-off will be spectacular! The sound and fury will be unbelievable! And the model will go out of sight overhead. This is the time that you should check for any possibility of grass fire around the launch pad if you are not using a tarp or launching from asphalt or concrete.

Most large model rockets will go higher than 1,000 feet and can top 3,000 feet easily. There is not a great deal of sense in trying for higher altitudes. Once you lose sight of the model, it is probably lost forever unless your tracking crews are using optical telescopes to follow it.

It may sound as though I am afraid of large model rockets. To some extent, I am because I have seen far too many of them turn in poor flights. They can become very hazardous if they are not handled and flown properly. Models powered by Type D motors are not as hairy as Type F-motored models—but then a Type F motor has four times the total impulse of a Type D.

You may have noticed that so far I have mentioned nothing about clustered motors. There is a reason for this. Some years ago clustered model rocket motors were the only available way to obtain higher thrusts and higher total impulses, just as clustered motors were once the only method available to NASA to obtain one million pounds of thrust for the Saturn-I. Today, however, the same objectives can be achieved with greater safety and much higher reliability by using a single large model rocket motor.

Figure 12-4: This large and impressive model was made by puppeteer Don Sahlin using the hollow log technique. Launch rods went up inside the body.

Clustered motors are difficult to ignite so that the model leaves the launch pad with all motors operating. They also weigh much more than a single large model rocket motor of equivalent total impulse.

Furthermore, clustered motors require models of greater diameter than a single-motored model, and thus they have higher base drag, pressure drag, and friction drag.

Anyone who tries to cluster Type F motors for fun is a fool. Nobody needs that much boost. If you think you do, you have not done a neat design job.

When flying large model rockets, one must keep clearly in mind the rules and regulations of the Federal Aviation Administration (FAA). The Federal Air Regulations Part 101, Subpart A, Subparts 101.1 a.3ii, a through d (Section 307, 72 Statute 749, 49 United States Code 1348) permits the flying of model rockets provided their gross weight does not exceed 16 ounces and they do not use more than 4 ounces of rocket propellant. Model rockets that exceed these limits are not considered to be model rockets by the FAA, so if you have a "biggy," you will be required to obtain in advance an Air Traffic Control clearance for *every flight.* Otherwise, you will be in big trouble with the feds.

You will also be in big trouble with nearly all the state governments and fire marshals because of the nearly universal adoption of the National Fire Protection Association's Code for Unmanned Rockets, NFPA Number 1122L. This suggested standard law, drawn up with the help of the National Association of Rocketry and the Hobby Industry Association, makes it a misdemeanor to launch a model rocket that exceeds the FAA limits.

You see, in model rocketry we have *voluntarily limited* the size and weight of the models we build and have cooperated with public safety officials.

There is more to it than that, however. There is financial and technical sense to these limitations. A high-performance Type F motor can cost as much as eight dollars. The model can cost upwards of fifteen dollars. When you start talking about big high-performance, high-altitude model rockets, you start talking about high-powered costs, too. Besides, a mere 4 ounces of rocket propellant will put a model rocket much higher than you can track, see, or recover.

Large model rockets are therefore not usually intended for high-altitude performance only. They are basically payload carriers, the weight-lifters of model rocketry. Although some payloads are quite small and compact these days, there are other types of payloads that cannot be miniaturized and lightened much—as we will now see.

Figure 12-5: Don Sahlin's big model lifts off under the thrust of high-impulse model rocket motors. The photo was made from a single frame of a high-speed 16mm color motion picture of the flight made by Sahlin.

Payloads

Many model rockets are built strictly for sport flying or contest flying. They carry no payload other than their own airframes. But one of the reasons for the existence of present-day rocket-powered vehicles is their ability to lift payloads to very high altitudes. In times past and even today rocket vehicles carry explosive warheads, signaling devices such as flares, rescue components such as ropes and lines, and scientific instruments. The most important of the payloads carried by rockets is people.

In model rocketry we do not work with explosive warheads. This is forbidden by the safety codes and by all the rules and laws regarding the sport. Explosive warheads are very dangerous, and direct handling of explosives is not part of the hobby of model rocketry. It takes a great deal of highly specialized training to be able to handle these materials with any degree of safety whatsoever. In addition, one must be completely familiar with fusing and arming procedures, something one learns only in the military services after many years. It is all far too hazardous for the average person to handle. If you are fascinated by explosives and things that go bang, join the armed services and get the proper training as an explosives and demolitions expert. Don't use model rocketry as your training ground.

In model rocketry we fly for fun and knowledge, not to conduct a small war!

Payload-carrying model rockets are the special province of the experienced, advanced model rocketeers. Many things have been done with model rockets, including pollution patrol, smog control

Figure 13-1: Payload-carrying model rockets can be designed to perform useful tasks. At the National Air and Space Museum of the Smithsonian Institution, Frederick C. Durant III, director of astronautics (left), holds a Cineroc, while Michael Collins, NASM director and *Apollo-11* astronaut (right), inspects a still-camera Cameroc.

studies, and other investigations relating to ecological concerns. And many more things can be done with payload-carrying model rockets that haven't been done yet. So it is still a wide-open field for careful research, creative development, and extensive testing—one more aspect of model rocketry that is full of challenge.

Model rockets are not usually designed to carry any old payload that comes along, although they can be and often are adapted to carry a wide variety of things. Usually a model rocket is designed around its intended payload, with the designer keeping in mind the size, shape, and weight of the payload plus the environmental factors that will affect the payload—acceleration, shock, vibration, heat, etc.

In nearly all cases a model rocket's payload is carried in the nose section of the model. It may be housed inside a hollow nose or placed just behind an aerodynamic nose shape in a cylindrical payload compartment that is structurally part of the nose assembly that comes off the model at ejection. By positioning the payload up near the nose, a better *CG-CP* relationship can be obtained. This can be important in many payload models because the additional weight of the payload usually causes lower lift-off accelerations and velocities, both of which require excellent stability characteristics in the model.

There are a number of payloads that are commonly carried in model rockets. They can be grouped into the following general classifications:
1. Passive, or dead load, payloads.
2. Optical payloads such as cameras.
3. Electronic payloads.
4. Active on-board payloads.
5. Biological payloads.
6. Special payloads.
Although there is some overlapping among these payload classifications, we can discuss them individually because their airframe and propulsion requirements are often very different.

Passive, or dead load, payloads

When the hobby of model rocketry began in 1957, the main emphasis was on propulsion and airframe technology because everything was new, and we early enthusiasts had to start with very simple models and very simple techniques. We did not have a large selection of model rocket motors from which to choose. In late 1957 and early 1958 we had only Type A4 motors to work with. We got Type B4 motors in late 1958. Therefore, we did not have the propulsion capability to lift large and heavy payloads. Transistors were just starting to become commercially available, and today's miniature cameras were years in the future.

The National Association of Rocketry recognized the basic payload-carrying ability of model rockets and developed what was then called the passive payload competition category. This is still the basic NAR payload category, which is also flown internationally under FAI rules. It is based on the ability of a model to carry one or more Standard Payloads to as high an altitude as possible with a given amount of propulsive total impulse.

The FAI-NAR Standard Payload is a small cylinder of lead with a diameter of 19.05 millimeters (0.75 inch) weighing no less than 28.35 grams (1.0 ounce). Simple, you say? Yes, once you have done it. But the rules go on to require that no holes be drilled in the Standard Payload; that it not be altered in any way; that the modeler be able to insert it into and remove it from the model at will; that the payload be totally enclosed within the airframe; and that the model be so designed that the payload cannot separate from the model in flight.

In essence, the model rocketeer is being put in the shoes of a real rocket design engineer who is told, "Here is the payload. It weighs X ounces and has these dimensions. You cannot change its shape. You cannot alter it in any way. You do not have to know what it is. Just build a rocket vehicle to take it as high as possible."

These passive payload specifications are not really very difficult to meet. Nearly any good model rocket can be converted into a passable payload model by the addition of a payload compartment or payload nose assembly. However, the design of a high-performance contest payload model is something else again because a converted sporting model usually does not have the optimum characteristics of a contest payloader.

Figure 13-2: The United States' first international winner, Talley Guill's gold medal contest payload model, which took first place at the First International Model Rocket Competition in Dubnica, Czechoslovakia, in 1966.

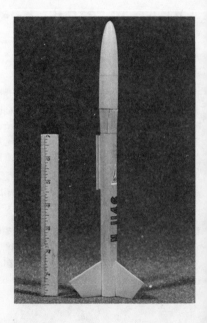

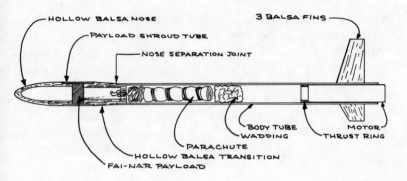

Figure 13-3: Cutaway view of a typical high-performance contest model built to carry the FAI-NAR payload weight.

Figure 13-2 shows J. Talley Guill's winning payload model from the First International Model Rocket Competition, held in Dubnica, Czechoslovakia, in May 1966. It is still a very good design in the PeeWee Payload and Single Payload events. These require a single Standard Payload and a Type A or Type B motor, respectively.

The cutaway of a typical Standard Payload model is shown in Figure 13-3. Note that standard off-the-shelf catalog-available parts are used throughout. Also note that there are little extra touches that contribute to the overall performance of the model because of optimization for one purpose only—carrying the Standard Payload to maximum possible altitude. The nose is hollowed out of balsa, or a hollow plastic nose is used. Every bit of weight that does not contribute to aerodynamic drag reduction, structural integrity, and payload retention has been eliminated or reduced.

The nose plug fits into the payload shroud with a very tight fit. See Figure 13-3. The best way to get this kind of fit is to wrap cellophane tape around the nose plug until it slides very tightly into the payload shroud. This is known as shimming. The nose plug's tight fit into the shroud prevents the payload from separating from the model.

I have built a lot of payload models for competition, and I cannot say for certain, "This is the very best payload model design possible." There are too many trade-offs. I do know that a short, squatty model will oscillate too much in flight, as we discussed in the chapter about stability. My best luck has come with long, slender payload models with a length-to-diameter ratio of 10 to 15. I also cannot tell you whether to use clipped-delta fins with 60 degrees of leading-edge sweepback, thus creating a fin that will not stall at low angles of attack but has lower normal force, or whether to use a high aspect-ratio tapered fin design that offers less induced drag but may stall at low angles of attack. Why the emphasis on angles of attack for the fins of a payload bird? Because the model, with its heavy payload nose, has the opportunity to swing to very high angles of attack before the fin can act to provide enough restoring force. Basically, the payload model has a far greater moment of inertia; once it starts to swing in pitch-yaw, it develops considerable momentum that the fins must overcome.

Payload contest models are fun. Now that we have new methods of altitude tracking, which will be discussed later, I am sure payload competition will regain the popularity it enjoyed in the 1960s.

Figure 13-4: The author prepares his contest payload model for its second-place flight at the First Internationals at Dubnica, Czechoslovakia.

Optical payloads

One of the most interesting model rocket payloads is a camera, and many model rocketeers have worked very hard to build and fly camera models. The first camera model on record was built and flown by Lewis Dewart, of Sunbury, Pennsylvania, in 1961. A small Japanese camera was simply strapped to the side of a model rocket. When the ejection charge popped the nose, it pulled a string that released the shutter and permitted the camera to take a photo of the ground below—or the sky and clouds, depending upon the direction the model was pointed.

Vernon D. Estes and Estes Industries, Inc., brought out the first commercial model rocket camera, the Cameroc, in 1965. The Cameroc allowed all model rocketeers to become in-flight photographers. The Cameroc lens points straight up through the tip of the nose. Therefore, the model must be over peak altitude and pointed down when the ejection charge goes off, ejecting the nose-camera and tripping the shutter. The Cameroc takes one black-and-white photograph per flight, hopefully while the nose is pointed toward the ground from a respectable altitude. The negative is a circle 1.5 inches in diameter. It is Tri-X film, which you can develop in your own darkroom (or even on the flying field) if you are a camera buff. If you are not, you can send the film to Estes for development. Don't take it to your local film processor because they do not have the facilities for developing circular negatives and because the ASA film speed of the Tri-X film must be pushed to ASA 1200 by special processing techniques. Standard processing won't work.

Other miniature cameras have been and can be developed for model rocket work. But the Estes Cameroc is the only model rocket still camera readily available at a reasonable price.

It also occurred to a number of model rocketeers that a motion picture camera in a model rocket would produce a spectacular piece of footage as the ground fell away and the model climbed to high altitudes. The first in this area was the movie camera model built and flown by Paul Hans and Don Scott, of Port Washington, New York, in 1962. This was a big model powered by a Type F motor because the smallest motion picture camera available at that time was the Bolsey B-8, a spring-wound 8-mm camera. It was heavy. Following months . of preparation, including flights of preliminary designs carrying dummy cameras, Hans and Scott committed their Bolsey B-8 to

Figure 13-5: The Estes Cameroc two-stage model carries a single-frame still camera in its nose. When the nose is popped off at ejection, a string is released to trip the shutter.

Figure 13-6: An Estes Cameroc took this photo of the prep and launch area at the Tenth National Model Rocket Championships at NASA Wallops Station, Virginia. Original negative is 1.5 inches in diameter.

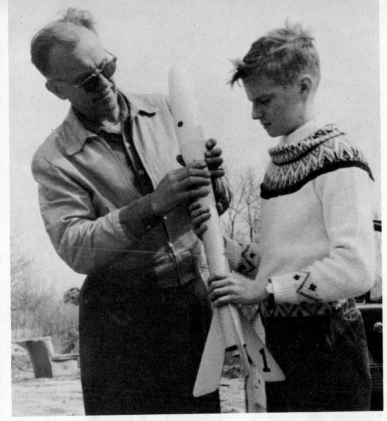

Figure 13-7: Charles and Paul Hans prepare the original movie camera model for its first flight.

flight. The lens looked out through a hole in the side of the nose section; the nose and body sections were recovered on separate brightly colored silk parachutes.

On the first flight everything worked perfectly. The model flew beautifully. Scott had to climb a tree to get the camera back. The color film was sent to the processing lab—and promptly disappeared! It was lost. The company replaced the film, but could not replace the flight footage. Undaunted, Hans and Scott tried again at the Fourth National Model Rocket Championships at the Air Force Academy in Colorado. This time the two modelers took the film to a different processing lab with very explicit instructions.

That first in-flight piece of color motion picture film was indeed spectacular. The boys sold it to Time-Life, Inc., who never used it but left it to languish in their voluminous files.

Vernon D. Estes and Estes Industries came to the rescue of the model rocketeer again. They hired Mike Dorffler, a young model rocketeer who had developed a very small and very lightweight

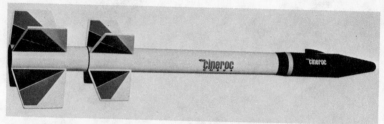

Figure 13-8: The Estes Cineroc model rocket uses Type D motors and carries in its nose a specially designed 8mm movie camera that looks back along the model body. Model is 1.6 inches in diameter.

model rocket movie camera. Dorffler's camera was refined and developed into the Estes Cineroc, one of the most elegant model rocket products ever to be put on the market. Fully loaded with its own cassette of Super-8 color film, the Cineroc weighs a mere 2 ounces (56.7 grams). It is battery-driven, has a 10-mm focal-length lens, shoots 31 frames per second at f:11 with a shutter speed of 1/500 to stop any rocket motion, and is 9.9 inches long and 1.75 inches in diameter. This tiny movie camera has taken some outstanding in-flight movies. It probably has thousands of other uses where a small, very lightweight movie camera is required.

If you want to fly cameras, I highly recommend the Estes Cineroc and Cameroc. They work, they are reasonably priced, and they can give you some spectacular results. Of all the model rocket payloads these two cameras are perhaps the most fun to experiment with.

Electronic payloads

Real research rockets flown by NASA and other agencies carry small radio transmitters to convert electrical signals from sensors—temperature and pressure pickups, for example—into radio signals that are sent to receiving stations on the ground. This gives scientists a moment-by-moment picture of what is happening up where the rocket is.

Model rocketeers have tried this, too. But it is no simple project, even today. You must have a thorough knowledge of electronics to get any meaningful information.

Figure 13-9: Frames from a flight of an Estes Cineroc movie camera model. The camera looks back along the body toward the fins. Sequence shows lift-off, arcing over at apogee, and ejection of parachute.

The first model rocket transmitter was designed and built by John S. Roe and Bill Robson, of Colorado Springs, Colorado; it was first flown in public at the Second National Model Rocket Championships near Colorado Springs in August 1960. About a year of work went into its design. The initial model was a simple transmitter whose schematic is shown in Figure 13-11. It is a small, lightweight

Figure 13-10: Bill Robson loads the first radio-carrying model rocket into a launch tower at the Second National Model Rocket Championships in 1960. Note the long antenna wire coming from the tip of the nose.

single-transistor radio-frequency oscillator broadcasting in the 27-megahertz Citizen's Band. No information is sent by this transmitter. It merely emits a signal that indicates, "Here I am!" By using a directional antenna on the ground, you can find the transmitter and the model rocket after it has landed.

The circuit was then modified to produce the schematic shown in Figure 13-12. The layout is shown in Figure 13-13. This unit was featured in the May 1962 issue of *Electronics Illustrated* magazine, which gives you some idea of how long the basic idea has been around. Roe added a pair of general-purpose transistors wired as a free-running multivibrator whose frequency is changed by a change in the resistance of a sensor. A typical sensor would be a photocell looking out through a hole in the side of the model.

Roe and Robson first flew their transmitter with such a photoconductive cell. When the model rolled in flight, the photocell reported every time the sun shone on it. This changed the audio frequency impressed on the r-f carrier.

Although you can build this transmitter if you wish, you can probably buy an Estes Rocketronics transmitter for about the same price. The basic Rocketronics Transroc operates on one of the twenty-three different Citizen's Band channels. There are Rocketronics accessory kits such as a microphone so you can listen in to the sounds of flight; a spin-rate kit with a photocell to determine roll rate; and a

B1: 9 VOLTS
C1: 3-30 MMF.
L1: 1 μh.
L2: 1.5 μh.
Q1: 2N1516 OR 2N384
R1: 220 KΩ, ½ w. 10%
XTAL: 26.97-27.27 MC.

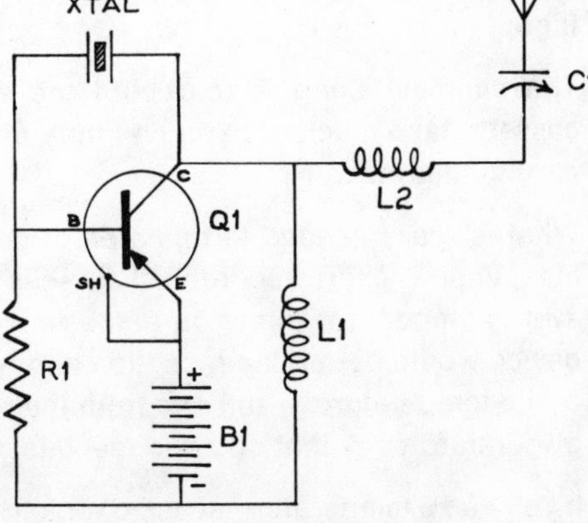

Figure 13-11: Schematic of simple model rocket radio transmitter.

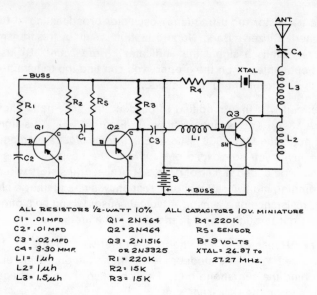

ALL RESISTORS ½-WATT 10%   ALL CAPACITORS 10V. MINIATURE

C1 = .01 MFD   Q1 = 2N464   R4 = 220K
C2 = .01 MFD   Q2 = 2N464   R5 = SENSOR
C3 = .02 MFD   Q3 = 2N1516   B = 9 VOLTS
C4 = 3-30 MMF.   OR 2N3325   XTAL = 26.97 To
L1 = 1 µh   R1 = 220K   27.27 MHz.
L2 = 1 µh   R2 = 15K
L3 = 1.5 µh   R3 = 15K

Figure 13-12: Schematic of single-channel AM-FM model rocket radio transmitter.

temperature-sensing kit so that you can measure the air temperature aloft.

Although many transmitters are now flying, the big problem still remains—devising a suitable ground receiving station with chart-recording capabilities. Ordinary CB receivers or walkie-talkies are suitable for receiving the radio signal from the transmitter in the model. But you must be able to do something with this brief and very transient information that comes back to earth from your model in flight.

The simplest thing is to record the audio signal on a portable cassette tape machine. You can then play it back at a later time for careful analysis.

What is really needed in this area of payload model rocketry is an inexpensive chart recorder that sells within the budget of the average model rocketeer or, at the very most, a school club. Such a device would permit the reception and recording of a chart or graph of the temperature or roll rate from the rocket-borne signal. It would give a hard copy that could immediately be deciphered.

If you had a temperature-sensing radio transmitter, a model rocket, a ground receiving station, and a chart recorder, you would be able to

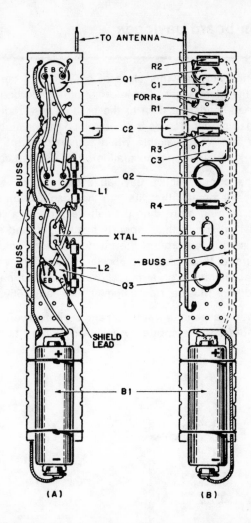

Figure 13-13: Parts layout of the single-channel AM-FM model rocket radio transmitter.

conduct valid air pollution studies with model rocketry. You would be able to detect temperature inversions, and you would be able to tell how high these inversions were by knowing how high the model went. By knowing how fast the model descends on a parachute, you could determine the average wind velocity aloft. Armed with this data, you could make highly scientific and meaningful air pollution forecasts. The full information on how to do this was published in an article of mine in the November 1972 issue of *Analog* magazine.

Active on-board payloads

The difference between this class of payload and the electronic payload is the place where the information is recorded. The electronic payload sends the data to the ground on a radio link. The active payload model records the data on board the model, and the information becomes available when you recover the model. To some extent, the camera models fall into this payload category.

However, the primary active payload used to date has been an on-board accelerometer that fits into the nose section of a model rocket. It is nothing more than a small weight with a scribe point attached to it, a spring that supports this weight, and some chart paper that is wrapped around the inside of the payload compartment. The first of these on-board accelerometers for measuring the maximum acceleration of a model was built and perfected by Lindsay Audin, of Hillside, New Jersey, in the early 1960s.

When the model accelerates at lift-off and during powered flight, the weight compresses the spring by a known amount that you have

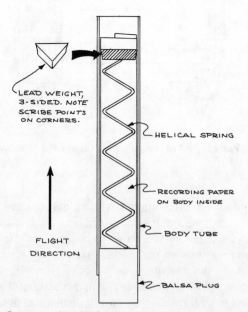

LEAD WEIGHT, 3-SIDED. NOTE SCRIBE POINTS ON CORNERS.

HELICAL SPRING

RECORDING PAPER ON BODY INSIDE

BODY TUBE

FLIGHT DIRECTION

BALSA PLUG

Figure 13-14: Cutaway sketch of an on-board accelerometer.

determined by ground calibration before the flight. The scribe on the side of the weight makes a mark on the paper inside the payload section tube. After recovery, you remove and unroll the paper. You can then check to see how far the weight compressed the spring because of acceleration. From this you can determine the maximum acceleration attained by the model.

There are probably many other types of on-board active payloads that could be built and proven. The biggest problem lies in keeping the weight low enough for a model rocket to easily lift it without seriously affecting the performance. There is a lot of room here for advanced work and inventive ingenuity. The surface has hardly been scratched!

Biological payloads

Sooner or later a model rocketeer gets the bright idea that it would be fun to fly a mouse. This is known as a Live Biological Payload, or LBP for short.

There have been a great many mice killed in model rockets. But a model rocket makes a very poor and very expensive mousetrap because you must first entice a mouse into the nose or the payload compartment. This isn't easy to do.

Unless you are conducting a valid scientific experiment under the supervision of a biology teacher, there is nothing that you can learn by flying an LBP. As a matter of fact, you can subject a biological specimen to the same environmental stresses on the ground without subjecting the animal to a rocket flight.

Thus, there is very little reason to fly any sort of LBP except to inflate your own ego. And there are many other things you can fly that are more ego-satisfying. That is why the National Association of Rocketry and the Hobby Industry Association frown on flying live animals.

The American Society for the Prevention of Cruelty to Animals (ASPCA) has even stronger objections to model rocketeers flying live animals. The ASPCA has completely shut down all model rocket activity in several schools and clubs by court order because live animal flights were reported in the news media.

There is a better and more fun way:

Special payloads

In 1962 Captain David Barr of the United States Air Force Academy proposed a great idea that was immediately adopted by model rocketeers. Captain Barr felt that an excellent test of a model rocketeer's ability would involve the flight and recovery of a *fresh* Grade A Large hen's egg without cracking the shell.

Egg lofting turned into a fantastically enjoyable area of payload model rocketry. It was and still is quite a challenge. A fresh Grade A Large hen's egg weighs an average of 2.7 ounces (76.5 grams) and has a minimum diameter of 1.75 inches (44.5 millimeters). Today's quality-controlled hatcheries mass produce eggs with very similar dimensions and weights. The eggs also have very thin shells, which adds to the challenge. An egg is a very fragile payload. Furthermore, if something goes wrong, you have a scrambled egg and not a dead animal.

The first successful egg flight was made by Don Scott and Paul Hans in the movie camera rocket at NARAM-4 in 1962. Since then, thousands of model rocketeers have flown egg payloads, and we have learned some interesting tricks from these flights.

First of all, an egg is very strong in its longest dimension and very weak across its minimum diameter. However, if it is completely and solidly cushioned all around, it will withstand a terrific beating. Today's model rocket egg crates, often referred to as hen grenades, usually use a low-drag, vacuum-formed plastic egg capsule made by Competition Model Rockets. This capsule does such a fine job of cradling and supporting an egg that I have had my CMR egg capsule accidentally come loose from its parachute, fall freely for 1,000 feet, land on packed soil, and not even crack the egg. To some extent, this has taken a bit of the fun out of egg lofting. Still, every once in a while there is a terrific boo-boo resulting in a nose-down prang with a fresh egg, and egg flies in all directions on the flying field!

Many American model rocketeers have become so proficient at egg lofting that they can now do it regularly with a single Type B6 motor. So the payload competition has been modified to add another category in which you must fly a fresh egg to as high an altitude as possible with limited total impulse—and get the egg back unbroken, of course, or face the humiliation of disqualification. These egg lofting competitions have become regular parts of each year's national meet. Model rocketeers, however, seem to have gotten these Mercury Class egg lofts well in hand.

As a result, in 1972 the NAR proposed dual egg lofting (Gemini Class) in which *two* eggs have to be flown together and recovered. Maximum altitude is scored. Disqualification results if either egg is cracked.

There is great excitement in egg flying, even when things don't work exactly right. It certainly separates the good model rocketeers from the balsa butchers. If you goof, you have a hilarious mess on your hands. If you succeed, you can cook your payload for supper. Egg lofting has all the thrill and fun of flying an LBP, without the headaches. Try it!

What else can be flown in a model rocket as a payload? Just about anything that weighs less than 6 ounces (170 grams) and will fit into any of the large variety of body tubes now available. There is not much to be gained by flying miniature kitchen sinks and other types

Figure 13-15: An egg-lofting model rocket using the CMR plastic egg capsule.

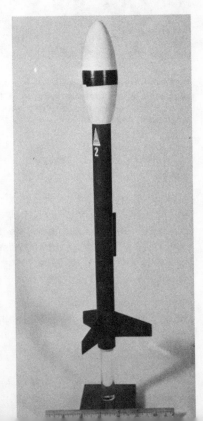

Figure 13-16: Apollo-8 Astronaut William Anders, one of the first men to reach the moon, looks on with delight as Michael Paskin holds aloft the fresh hen's egg he flew and recovered intact at NASA Wallops Station, Virginia, in 1964.

of goofy payloads. But many things could be flown in a model rocket that have not yet been aloft under rocket power.

Payload-carrying model rockets present some interesting design and construction problems. They can be very complicated and very expensive, depending upon the type of payload carried and the performance desired. In many cases the model rocket is merely the inexpensive carrier vehicle for a very expensive payload. This is where the reliability of model rocketry really pays off.

Payload-carrying model rockets are another province of the advanced model rocketeer and another proof of the statement: Model rocketry is fun!

Scale Models

Since model rocketry is space rocketry in miniature, it is only natural that model rocketeers would want to make their models resemble the "big ones" as closely as possible. The construction and flight of exact miniature replicas of full-sized rocket vehicles and space vehicles is model rocketry at its very best. It is the province of the true craftsman.

One of the first model rockets ever built was a scale model, a miniature version of the United States Navy's Pogo-Hi parachute target rocket. The model builder was Chuck Moser, an engineer on the Pogo-Hi project at the Physical Sciences Laboratory of New Mexico State University. He completed his model in March 1957.

Scale model rocketry combines craftsmanship with research. It reaches its pinnacle when a miniature replica, complete and correct down to the smallest details of markings and coloring, thunders off the launch pad for a straight flight and a perfect recovery. Months of research and work have probably gone into the model. Often such models are so beautiful that you may hate to fly them. But having a beautiful scale model that has actually flown is much more satisfying than having one that sits forever earthbound on a shelf without ever being borne aloft under rocket power.

Scale model rockets have been built with such fine workmanship and attention to details that they qualify in all respects as museum scale models while also being capable of flight. Some model rocketeers even duplicate the launching facilities and equipment so that their scale model takes off from a scale launch pad.

Many flying scale model rockets reside in the National Air and Space Museum of the Smithsonian Institution. In some cases they are the only models of a particular vehicle in existence to help trace mankind's steps to the stars. As a result, they have become valuable additions to the national collections. Under the direction of Frederick C. Durant III, the NASM Astronautics Department has always been sympathetic to scale model rocketeers, some of whom are among the best astro-historians alive today.

Scale model rocketry is not something that can be done overnight. The creation of a good scale model often takes weeks or months. It may take a long time simply to acquire the information that will ensure that you really have a scale replica. And construction cannot be rushed. If you hurry, you are likely to put that last frantic coat of paint on a perfectly built model only to have it blush or run, spoiling weeks of work. Do not start a scale model the night before a contest. Scale model rocketry takes weeks or months of planning, thought, patience, careful workmanship, attention to the smallest detail, and a lot of flight experience with nonscale sporting and contest models. Unlike participants in many other flying sports, scale model rocketeers are, in general, top-notch fliers as well.

Figure 14-1: The epitome of model rocketry is the construction and launch of a scale model, duplicating in miniature a full-sized rocket vehicle that flies.

The rewards of scale modeling are great. There is an indescribable sense of pride and accomplishment when your model is placed on display after a flight. In contests there are trophies and prizes to be won. In national and international meets there are always events for scale models because scale modeling brings out the best craftsmen, designers, researchers, and contest fliers.

Once you have become a scale model rocketeer, all other model rockets seem to be just paper and balsa look-alikes.

The challenge of scale modeling never ends. When you think you have done everything, you will always find another scale model beckoning to you. No modeler has ever built the perfect scale model. There are always improvements to be made, little flaws to be corrected on "the next one," or there are new techniques to try. New full-sized prototypes are always being developed for duplicating in miniature, and some of them pose real challenges to one's modeling abilities.

By participating in scale model rocketry you are likely to become such an expert on one particular space rocket or vehicle that you could talk intelligently with the project manager himself if you had to. And you will discover that you have become hooked on the pursuit of excellence.

Important points

There are several important points or steps that you should follow to build a good scale model. Occasionally you will discover an exception to these rules, but they have come from years of scale model work and over a decade of experience in local, national, and international contests. Even if you do not wish to fly your scale model in contests, these pointers will serve you well and help you to achieve a perfect scale model of that vehicle you are so fond of.
1. Select a good subject to model.
2. Get adequate scale data before you start to build.
3. Using your data and the model rocket manufacturers' catalogs, select the correct size for your model, using as many commercial parts as possible; also select the type of model rocket motor that you will use to propel your model.

4. Prepare accurate working drawings and make accurate calculations of estimated *CG*, *CP*, weights, and flight performance.

5. Build a less-than-perfect flight test model first to ensure that your later highly detailed scale model will perform as you want it to.

6. Build two or more detailed scale models of the same prototype side-by-side at the same time.

7. Take your time and do your best work. Don't take shortcuts.

8. Flight test your less-than-perfect model and your detailed scale model.

9. Keep building progressively better scale models of the same prototype.

Selecting a scale model

The most critical and important step in scale modeling is the selection of the proper prototype to model. If you do not do this carefully, you may have considerable trouble in building the model or getting it to fly. Too many modelers have become discouraged

Figure 14-2: Some scale models are so accurate that they have been put on museum display. Here Leroy Piester of Centuri Engineering Company (right) presents a 1:45 model of the Apollo Little Joe II to Frederick C. Durant III (left), director of astronautics at the National Air and Space Museum of the Smithsonian Institution.

about scale modeling because they made the wrong choice of a prototype. As a result, they had difficulty with their project.

The first step in the selection process is an honest self-evaluation. You must honestly evaluate your own abilities. If you are not very good at construction, assembly, painting, or other workmanship factors, you should not select a model to build that is beyond your abilities in that one particular weak area. Incidentally, this holds true for all model rocket activities, but it is "more so" in scale modeling. You should not select a complex model if you do not think you can finish it; this is not copping out on a challenge, but simply being mature enough to recognize your own shortcomings and strengths at that particular stage in your development as a model rocketeer.

Nobody is going to fault you for starting out in a simple manner. In fact, it is smart to begin with a simple model that will result in a good, reliable, flyable scale bird. Do not let your personal enthusiasm for a certain full-sized rocket vehicle get the upper hand over common sense. You may be very eager to build a Saturn-Ib, a Titan-IIIC, or a Space Shuttle because the real one is in the news or has made history. Or you may have been able to get all sorts of data on the Nike-Hercules that is on display in the city park or the Atlas-D that is standing at the entrance to the nearby Air Force base. You may even have been able to get near a real rocket vehicle to photograph it from all angles and to measure it carefully. But do you really have the ability and experience to build such a complex, high-detailed scale model on your first attempt?

The best model to start with is a simple, straightforward scaler that looks much like the sporting model rockets you've been building. Your first scaler should be a single-staged vehicle with a cylindrical body, a simple ogive nose, and plenty of fin area. The color scheme should be simple, and it should not have a great deal of detail.

Another important selection factor, which we will discuss later in greater detail, is the availability of information on the prototype you wish to model. Scale information on current military rockets is usually very difficult, if not impossible, to obtain because of military security precautions. This is particularly true of recent guided missiles. In fact, precise dimensions and shapes of some military rockets may remain classified for as long as twenty years, even though there are hundreds of them on public display all over the world!

On the other hand, if you have chosen a very early rocket or space vehicle, you may have great difficulty in locating information be-

Figure 14-3: A photograph of the real I.Q.S.Y. Tomahawk as launched at NASA Wallops Station, Virginia.

Figure 14-4: Connie Stine hooks up the igniter on her scale model I.Q.S.Y. Tomahawk. She also built the scale launcher.

264

cause pictures and drawings are often destroyed to clean out the files and make room for more paperwork. Accurate information on many early rockets is difficult to obtain even if the rocket was well-known. For example, I have literally traveled all over the United States and Europe to obtain precise data on the old German V-2 (A4) rocket—photographing one from this angle at White Sands, getting that detail in Munich, learning about another aspect in the London Science Museum, etc. If you have chosen a rare bird, you may well have to work for several years to obtain information!

For those modelers who are starting in scale modeling, I have found it best to begin with a kit, and there are numerous highly accurate scale model kits available from several manufacturers as of 1975. If built to sound scale standards, these kits will give you a good taste of scalers, will start you toward learning the competence in construction and finishing required for scale models, and will produce for you a good-looking scaler that is capable of holding its own in most competitions.

There are two scale model kits now on the market that I consider from experience to be a beginner's best introduction to scale modeling.

The Centuri I.Q.S.Y. Tomahawk was specifically developed by me as a beginner's scaler. Scale drawings for the model are presented in Figure 14-5. In the kit model the plastic nose is not precisely to scale, and you should join the two body tube sections and fill in the joint to make one long body tube. Otherwise, this is an excellent bird with which to begin scale work.

The Estes K-51 Sandhawk has many highly detailed plastic parts, making assembly seem much easier, though it really isn't by the time you clean up the plastic joints. The Sandhawk is a big bird with lots of detail on it.

Other kits that are good for beginners are: Estes K-26 ARCAS, Centuri KS-15 Nike-Smoke, and Centuri KS-1 Mercury-Redstone.

If you disdain kits and ready-made parts, you can try your hand at scratch building, literally starting from scratch. Some excellent prototypes to model include: United States Navy Viking #7 or Viking #10, United Kingdom Skylark, French C.N.E.S. Veronique, Sandia/NASA Nike-Tomahawk, United States Navy Pogo-Hi, United States Army Honest John, and United States Army Jupiter-C.

In selecting your prototype, you might want to consider these tips on certain vehicles. The Aerobee series should be avoided because of

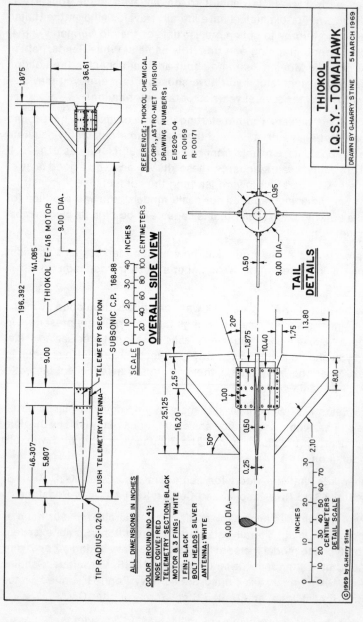

Figure 14-5: Scale drawing of the I.Q.S.Y. Tomahawk.

the booster rocket, which is attached below the vehicle by an open tubular framework; this is very difficult to model well. Multistaged vehicles should not be attempted at first, even if the upper stages are inoperable dummies. Air-launched missiles such as the United States Air Force Falcon series and the United States Navy Bullpup and Sparrow should be avoided because they have guidance control fins up near the nose and were designed to be launched from airplanes at high initial airspeeds. Antiaircraft rockets and similar vehicles typified by the United States Army Nike series, the United States Navy Terrier, Tartar, and Talos series, and the Russian SAM missiles should be tried only by experienced modelers. The Thor, Jupiter, Minuteman, Atlas, and Titan missiles are finless and there-fore require the addition of clear plastic fins to make them aerody-namically stable as model rockets; they too should be attempted only by experienced modelers.

Even with these restrictions, there are hundreds of rocket vehicles suitable for scale modeling, and new ones turn up all the time.

Obtaining scale data

Once you have chosen a prototype to model—and you may be smart to choose more than one in case you run into difficulty at some later stage—the next step is to get information that will permit you to build a true scale replica and not just a semiscale bird that looks something like the real one.

You may already have done some research in the local library, which may have books with pictures and drawings. The National Associa-tion of Rocketry has a growing series of scale plans available to members (another enticement to join). A good source is a fellow modeler who may have collected photographs and drawings. As a matter of fact, a great deal of scale data swapping goes on among scale modelers.

However, to get good data, you should attempt to get to the source—the manufacturer who made the real rocket vehicle. If you don't know who made it, find out who uses it, such as NASA. Then send your first letter there.

Your letters should be typed neatly on clean white paper or written legibly in ink on good stationery. Don't scrawl a penciled note on a scrap of notebook paper. Include your complete return address, including ZIP code. The results you get may depend upon the appearance, neatness, correct spelling, and correct grammar of your letter.

Don't concoct a fake "aerospace research center" letterhead or give yourself a fancy title. That doesn't help at all. Everyone in the aerospace business knows everyone else and also knows what is going on; a phony letterhead will be spotted immediately and may be chucked into the circular file along with crank mail.

Your letter should state clearly why you want the information, what vehicle you want information about, and the exact nature of the data requested—photograph, dimensioned drawing, etc. Be specific. Don't ask for everything they have. You won't get it even if they're going to throw it away. Remember, no aerospace company or government agency is going to turn inside out for you, taxpayer or youth science education notwithstanding. They will perhaps send you whatever they are able to lay their hands on in a five-minute scan of the public affairs office, which is where your letter is going to end up because nobody knows what else to do with it. Besides, they are always pretty busy or short of money this year. I am not trying to deride government agencies or aerospace manufacturers; I'm only trying to explain their possible behavior toward you. Just put yourself in their shoes, trying to answer dozens of letters like yours every day! Some agencies and companies are very good at sending data, having anticipated your request and gotten the information together for you and all the others who write in. It may take several weeks to get an answer. If you don't get a reply in four weeks, write another letter, and please be polite.

Figure 14-6: The Estes Sandhawk kit will make a fine scale model for getting started in scale work.

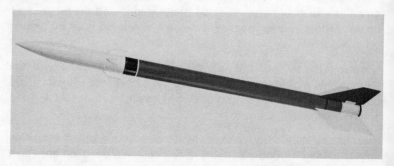

The results of your initial request may produce nothing more than a full-color brochure that is beautiful and expensive, but contains very little scale information of use to you. Write again, thanking them for the brochure and asking for specific information, explaining why you need it to build your scale model. Above all, don't give up! Persistent model rocketeers who were polite have actually succeeded in getting valuable historical information declassified or rescued from oblivion, thereby performing a highly commendable service to aerospace history.

Remember, it is quality, not quantity, of information that you want. Don't ask for the entire stack of factory drawings. That may amount to more than fifty pounds of drawings, including details of every little internal nut and bolt. This is what is known as wall to wall data, most of which is pretty useless when you sort it out.

Minimum scale data

For competition purposes the minimum amount of scale data that you must present to substantiate your claim to scale under the requirements of the NAR is: overall length, diameter, nose length, fin length, fin width, length of transition pieces, fin thickness, color pattern, and at least one clear photograph. More data are certainly desirable.

You should begin to collect and file scale data because it soon begins to grow and become exceedingly valuable to you. Make a file folder or get a large manila envelope to hold all of the information on a given prototype. Clip magazines. Use copying machines to get copies of the data you want. Scale data need not be originals.

It is sometimes possible to measure, or tape out, a prototype if you can get near the real thing. If you do this, you should make careful notes as you go along, and date the notes. A camera is handy for recording facts that you will later forget and for recording color data. You might be surprised at the number and variety of rocket vehicles that are on display and just waiting to be taped out!

Scale data can sometimes be obtained from a professional or museum model, but don't count on its being accurate unless you can back it up from an independent source. Some models are not accurate! In spite of advertising claims, very few nonflying plastic kit models are accurate. Remember that the museum model builder or the plastic kit manufacturer has been faced with the same problems you have in acquiring accurate scale information, and he has often proceeded with incomplete or inaccurate data because of the pressure of deadlines.

As you continue to collect scale data, you will discover that you cannot trust some of it! Data from different sources may not agree. You may have to do some careful research to determine which source is correct. You will run into a problem long-known in the history of scale modeling as The Problem of the Lost Inch. A mistake often is transferred from one document to the next for years and years before somebody, usually a scale buff, finally spots it.

You should build a model of a particular prototype with a specific serial number, paint pattern, etc., unless the vehicle was manufactured by the thousands on an assembly line with little or nothing except a serial number to differentiate between individual vehicles. If not mass produced, make a model of a particular vehicle, for variations within a type may be great. For example, every one of the fourteen United States Navy Viking rocketsondes launched was different. Over 65% of the German V-2 (A4) rockets launched at White Sands had major external changes in their basic configuration, and every one of them had a different paint pattern! The Mercury capsule Freedom-7 was different from Liberty Bell-7, and if

Figure 14-7: Another good scale model for beginners is the Centuri Nike-Smoke scale model kit.

you build a Mercury-Redstone, you'd better have the proper Mercury capsule atop the properly marked Redstone. Most of the Saturn-I vehicles were different, and the paint pattern usually shown for the Saturn-V is not the one that was used for the actual flight vehicles. If you get involved with a vehicle such as the Delta, where over one hundred of them were launched and no two were alike, you have lots of research to do!

Designing a scale model

When you have collected enough data to get started, you are ready to design your scale model. You should be completely familiar with the commercial model rocket parts that are available so that you can use them where possible. You should know what model rocket motors are available and what their performance characteristics are so that you can choose the proper power plant for the model. With all of this in mind, you can begin to size your scale bird.

Sizing is a very important step. If you have chosen a prototype with lots of external details and many complex shapes, you should build a big scale model. Little details are often exceedingly difficult to make and attach to a very small model. And on a small model details seem to get lost because the human eye cannot see them so well. On the other hand, if you have selected a rather simple prototype without many details—basically just a nose, body, and fins—your model will

Figure 14-8: Sometimes building a scale model in different scales with body tubes of different diameters will help you to get a good scale model. The author built these four I.Q.S.Y. Tomahawks in different scales to determine the best size.

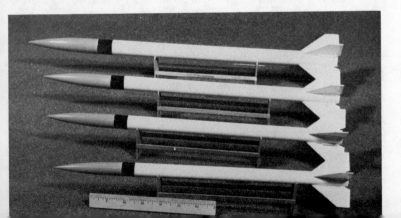

look better if it is smaller. In scale modeling, the size of the model is not so important as its overall appearance and flight characteristics. Large models are usually impressive and will often garner more scoring points from inexperienced scale judges, but you have many big problems in building and flying a big scaler!

You may also have to size your scaler on the basis of what you intend to do with it. If it is a pure competition scale bird, you don't have many restrictions in this regard. If you want a scale model that is also good for altitude, you will look for low total drag, high impulse-to-weight ratio, and a size and color scheme that will enable the model to be seen by the altitude tracking crews.

Again, use as many commercial parts as you can. You will have to make plenty of custom parts as it is, so don't make things too difficult for yourself. Use commercial body tubes. Many commercial noses and transitions can be used or modified for scale models, too.

The biggest single problem of most scale models is their high weight. The majority of scale modelers usually end up with a model that weighs too much and therefore requires a big motor to get it into the air. As a result, some scalers turn in very hairy flights because they are underpowered and just barely manage to stagger into the air and get high enough to crash.

So choose a motor for your scaler before you start to design and build it. Match the motor to the model just as you would if you were designing and building a sport model. If your scaler is underpowered, it is likely to prang. If it is overpowered, you will probably lose it on the first flight.

There are no hard-and-fast rules for sizing a scale model because each case is different. Sometimes you may have to do a little experimenting by building two or more semiscale flight test models of the same prototype in different sizes to find out which one suits your needs and desires as well as which one flies best. I have built the I.Q.S.Y. Tomahawk in four different scales with four different body tube diameters, just to see which one flew best. I have also built the ASP-I in three different sizes for three different types of motors.

To design your scale model properly, you will have to do a little drafting. This is not difficult, and you should learn how to do it because it will help you with your other model rocketry activities. Today, drafting equipment is readily available at reasonable prices. You will need a good drawing board, a T square, a couple of triangles, a good pencil with 2-H lead, an eraser, an engineer's triangular scale, a French curve or two, a drawing compass, and a good protractor. Get a little practice first by designing and drawing a couple of sport models. You will be surprised to discover how easy it is to do model rocket drafting!

Having decided on the size of your scale model, you must now prepare a full-sized working drawing of the model so that you can determine the sizes and shapes of various parts. To do this, you must determine the scale of your model.

Scaling a model

Scale refers to the relationship of a model's size to its full-sized prototype. The most common method for determining scale is to compare the diameter of the prototype to the diameter of the model. The scale of the model is the ratio between a dimension on the model and the same dimension on the prototype.

For example, if the diameter of the prototype is 31.0 inches, and you plan to use a body tube with an outside diameter of 1.04 inches, you divide 31.0 by 1.04. This gives you the number 29.81. The scale of your model will then be 1 to 29.81, written 1:29.81. It means that 1 inch on the model is equal to 29.81 inches on the real vehicle.

Using a piece of paper and a slide rule, calculator, or ordinary long division, divide every dimension of the prototype by the scale ratio—29.81 in our example. This will give you the dimensions of every part of your model.

Naturally, a 30-degree angle on the prototype is still a 30-degree angle on the model because an angle is not a dimension that changes with model size. So do not divide angles by the scale ratio—or you will end up with a wild-looking model!

Using the dimensions you have calculated, draw a full-sized working plan of your scale model. If you are lazy, you can take your prototype data to a blueprint shop and have them make a photographic reduction or enlargement of the prototype plans to exactly the size you want your model. This saves you from having to do the drafting, but it may cost a bit more.

From the working drawing that you now have, you can determine the size of the parts. By tracing the fin outline onto a piece of stiff paper

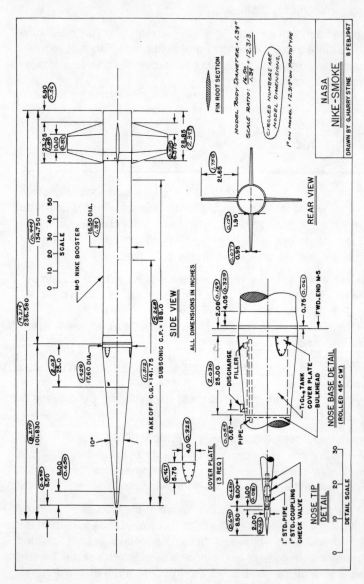

Figure 14-9: Scale model drawing with both prototype dimensions and model dimensions scaled properly.

or cardboard, you can make a template for cutting out all of the fins to the same size and shape. You can also make paper templates for the nose shape, transition shape, boattail shape, and other shapes on the model; this helps to assure that you shape them correctly.

But don't start to build yet! If you really want a successful scaler, stay on the drawing board with the paperwork and resist the temptation to cut balsa until you have figured that model right down to the last possible calculation.

Calculate the *CP*, using the Barrowman Method. It is not a good idea to use a cardboard cutout or the swing test for a scale model because most scalers are a bit shy on fin area. If you use the cutout method, you may end up having to put too much weight in the nose, thereby reducing the performance of the model unnecessarily. The same holds true for the swing test, which you cannot conduct until after you have completed the model, and then it may be too late to take proper corrective action on *CG-CP* locations. If you use the Barrowman Method, you will obtain the correct *CP* location and a better scale model because you will be able to locate the *CG* and determine the stability with a greater degree of confidence.

Make a trial run at estimating the *CG*, using the weight of the various parts and their distance from a common point such as the nose tip. Basically, this is the same procedure that you go through in calculating the *CP* of the model by the Barrowman Method, except that you use the weight instead of the area.

Also calculate the flight performance, using Malewicki's methods in Centuri TIR-100, making several estimates of the drag coefficient C_d to bracket the altitude range in which your model will be flying. This

Figure 14-10: Scale model rocketry knows no country. The Polish national scale model team at the First World Championships for Space Models in 1972 flew (left to right) a French Diamant, a USA Saturn IV, and a USSR Soyuz.

will also help you check your choice of motor. Sometimes, if you are lucky, it is possible to obtain data on the actual subsonic drag coefficient of the prototype, and this results in extremely simple and accurate performance calculations.

Why all this paperwork? So that you can discover ahead of time whether your model would be too heavy, underpowered, or require too much nose weight. You may also discover that if you built it, it would not fly in a stable condition no matter what you did. You may even discover that it would be impossible to build the model in the first place! In any case, time spent over the drawing board always pays off in scale model rocketry just as it does in full-scale rocketry, and for the same reasons. This is particularly true if problems happen to show up in the design process. It doesn't take much time to do a drawing or to make calculations; it takes far more time and money to correct a mistake after you've built the model!

Recheck your numbers—and then believe in them and rely on them. Once I spent several weeks building a scale bird. I calculated the flight time from burnout to impact at 6 seconds total. At that time I could not get a Type D motor with less than 6 seconds of time delay, and I had to have a Type D motor to get the bird off the ground. The model was too small for a Type E motor. So what could I do? I tried to beat the odds involved with slightly rounded-off numbers, slide rule inaccuracies, and production tolerances on motors, hoping that I might end up on the favorable side of the tolerances. I didn't. The model popped its 'chute a fraction of a second after the nose entered the region of exceptionally high drag coefficient known as the ground.

Building a scale model

There aren't many special tips on building a good scale model if you have done everything correctly up to this point. Just take care and do good work.

As mentioned earlier, it is often wise to build a less-than-perfect semiscale flight test model first. Do a good job, but don't super-detail it. If you get a run in the paint job, forget it. This is strictly a test bird. Build it and use it to check your stability and performance calculations. Go out and fly it to see if it performs the way the

Figure 14-11: George Pantalos with his Thrust-Augmented Long-Tank Delta scale model at Vrsac, Yugoslavia.

numbers say it should. You may run into little problems with this flight test model that you can correct before going on to the fully detailed scaler.

For example, my first Nike-Smoke scaler checked out fine on the board, but I ran into the rare dynamic stability problem known as pitch-roll coupling. The model would go progressively more unstable as the thrust phase progressed, finally spinning around and around in a flat spin 50 feet up at burnout. Two nose weights were required to correct the difficulty, and the model still went awry every once in a while. The problem was cured once and for all by building a larger model of the Nike-Smoke.

When you finally sit down to build the super-detailed scale model, build two of them at the same time. It doesn't take much more time and effort to make a pair, and it will give you a spare, just in case. Somebody might sit on your scaler two minutes before you get ready to fly it in that regional contest. You might even go back at this point and bring your less-than-perfect semiscale flight test model up to full scaler standards.

Figure 14-12: Otakar Saffek of Czechoslovakia, the first World Champion scale modeler, with one of his early scratch-built 1:100 scale models of the USA Saturn V.

Improving your scale model

Most of the NAR National Champions and FAI World Champions and medalists have learned that they can improve their scale models by building the same scaler over and over again in increasingly improved models. You learn little tricks and shortcuts every time you build a new model of the same design. You improve your finishing, detailing, and workmanship with each new model. And not only do the models get better, they go together quicker. This is what is known as progressing on the learning curve.

J. Talley Guill, many times United States Junior and Leader National Champion and an FAI gold medalist, built models of the USAF Convair MX-774 HiRoc for over five years, finally perfecting them to the point where they were outstanding scale models that would fly precisely as he wanted them to. His father, A. W. Guill, had the same thing going for the Astrobee-1500. Charles Duelfer built models of the USAF GAR-11 Falcon from 1963 to 1968. Howard Kuhn's Argo D-4 Javelins are works of art. Otakar Saffek of Czechoslovakia, 1972 and 1974 World Champion scale modeler, built over a dozen models of the Saturn-V in 1:100 scale, improving every time. His final achievement was perfect down to scale corrugations and took him 2,000 hours of work.

Once you get a good combination and a good scale, stick with it. Improve it. Do research on additional details. Lighten the model here. Strengthen it where necessary. Put a better finish on the next one. Strive for high reliability in flight.

When you get tired of modeling one prototype, put the scaler away for a few months. Come back to it after you have built others, perhaps of a more complex nature.

When you've grown tired of building the flying part of the rocket vehicle system, tackle a scale launching complex! These are undoubtedly the most impressive scale model systems in the world. They take months and often years of work to achieve. Some modelers motorize various parts of the launch complexes, creating scale launch rails that move up and down and turn in azimuth, just like the real ones. Others have service towers that move away under remote control and umbilical towers that swing away just before ignition.

In scale modeling there is literally no end to what you can do. It can give you a lifetime of enjoyment and challenge. When you are a good scale modeler, you are a member of a very small and highly select clan of elite model rocketeers.

We need more elite model rocketeers. Come on in! There is plenty of room for more!

Altitude Determination

Earlier we discussed methods of calculating altitude performance of model rockets and learned that one cannot accurately determine achieved altitude by computation alone because of all the variables and parameters we are unable to control or to measure accurately. We also saw that determining achieved altitude by timing the flight of the model to apogee was extremely inaccurate, too.

At the suggestion of Douglas J. Malewicki, then at Cessna, and Larry Brown, then at Centuri Engineering Company, Bill Stine decided to look into the possibility of determining altitude by timing the fall of a standard marker streamer with a fixed set of dimensions and a fixed weight. The theory behind this is that the marker streamer—a piece of 0.0001-inch thick polyethylene film 1 inch wide and 12 inches long with a 3-gram nose weight taped to one end—would fall at a constant rate of speed. It was to be packed atop the recovery device in a model and ejected with the device, hopefully at or near apogee. The time required for the marker streamer to reach the ground from the moment of ejection would be equivalent to the altitude. Naturally, this method would determine the altitude of the model at ejection, not necessarily at peak altitude. For sporting contest work every model would carry the standard marker streamer, and the time of fall of each streamer would be scored. The model whose marker streamer took the longest time to reach the ground obviously ejected the marker at the highest altitude and therefore was the winner.

Some interesting facts were discovered when Bill flew some experimental shots. Larry Brown had determined that the standard marker streamer would drop at a constant rate of 18 feet per second; he made a series of test drops from a known height on the Phoenix fire department's training tower. Although Bill recorded a spread in marker drop time of as much as 15% in identical models powered by motors from the same production lot, he came to the conclusion that what he was recording was really not a 15% variation in altitude, but a variation in motor total impulse. Remember that the burnout velocity is a function of the total impulse, and that the maximum altitude is a function of the square of the burnout velocity. Therefore, a very small variation in total impulse would produce a large variation in altitude.

Figure 15-1: Model rockets can be tracked in flight with very simple equipment to obtain a definite figure for the achieved altitude.

It was not possible during the early tests of this altitude determination method to determine whether or not the fall time of the marker streamer provided a good piece of altitude data. Fortunately, even though the marker streamer method is excellent for conducting unofficial altitude contests without a great deal of equipment, we have other methods of determining with great accuracy the altitude achieved by model rockets.

When a big rocket flies at White Sands or Cape Canaveral, its flight performance and achieved altitude are determined by tracking the vehicle in flight with electronic devices like radar or with optical devices like telescopes or laser rangers. Such tracking devices provide reasonably accurate data on the vehicle's position in space at any given instant.

In model rocketry tracking is also used. However, electronic tracking of model rockets has not been employed to date because of the very high cost of the equipment and the very small size of the models. Radar tracking of model rockets has been done using NASA radars at Wallops Flight Research Center, but such tracking appears at this time to be impractical because of the high cost and complexity of radar. In addition, model rockets present a very small radar target. (Yes, they can be tracked by radar, even though they are nonmetallic. All that really provides the signal reflection is an object with a different dielectric constant than air.)

Therefore, optical tracking is almost universally used in model rocketry.

Optical tracking is just a fancy term for following the model in flight by eye and aiming some sort of measuring device at it. So far, the state of the art in laser tracking puts this tracking method in the same league with radar, so optical tracking with "calibrated eyeballs" assisted by tracking theodolites is the method of the moment.

Without a great deal of expense and knowledge, simple and reasonably accurate optical tracking equipment can be built by almost anyone with access to a junior high school shop or, in some cases, a well-equipped home workshop.

Optical tracking provides one and only one piece of information, but it is an important piece. It is a figure for maximum altitude achieved. This information is useful in design evaluation, staging studies, and competition. Optical tracking combined with simple trigonometry (simple because you do not have to understand trig to work with it) has provided an easy, accurate means of obtaining this information since 1958.

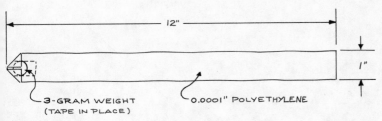

Figure 15-2: Sketch of the Standard Altitude Marker Streamer developed by Bill Stine.

There are two basic optical tracking methods—one a cheap-and-dirty method that will give a general figure for achieved altitude, and the other a sophisticated and highly accurate system that has been used worldwide for contests and record-setting purposes. There is also a third system that has been tried and proven in the field, but has not seen wide application because very few people know about it.

The cheap-and-dirty system uses a tracking device that measures the elevation angle of the model as seen from a single tracking station a known distance away from the launch pad. Two people are required to run this system—one to launch the model, and the other to track it in flight.

An elevation-only tracking device is shown in Figure 15-3. This is the simplest form of an elevation-only tracker. Others can be worked

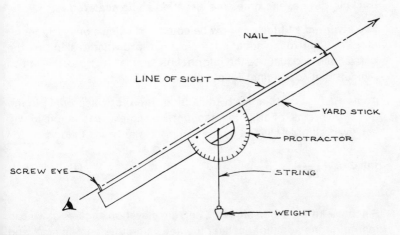

Figure 15-3: Simple elevation-only tracking device.

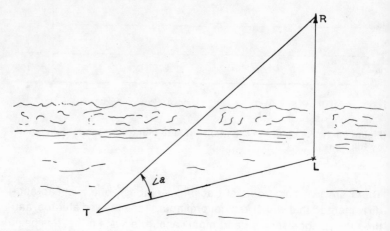

Figure 15-4: The geometry of a single-station, elevation-only altitude tracking system.

out. The setup for its use is shown in Figure 15-4. The launch pad is at *L*, the tracking station at *T*. The distance *LT* is measured beforehand and is usually on the order of 300 feet (91.4 meters) or more—and the more the better because the greater the distance *LT*, the better the overall accuracy of the system. When the model lifts off from *L*, it is assumed that it flies vertically or that the distance from a point on the ground directly under the model to the tracking station does not change from the original *LT* distance. The tracking station operator follows the model to the peak of its flight, locks his tracking device, and notes the elevation angle achieved.

The achieved altitude can now be computed using simple trigonometry. We'll derive the equations here, but if you want to skip over the derivations and just use the information, that is up to you. It is not so difficult that you cannot follow the derivation; others have.

Angle *RLT* is a right angle by definition. Therefore, according to the basic theorems of trig, the tangent of angle ∠ *a* is equal to the achieved altitude divided by the ground distance, as follows:

$$\tan \angle a = \frac{RL}{LT} \tag{1}$$

Since we know the distance *LT* and the elevation angle ∠ *a*, we can rearrange the equation, collect the known factors on one side, and come up with:

$$RL = LT \times \tan \angle a \qquad\qquad\qquad (2)$$

The tangent of the elevation angle can be found by looking in a tangent table such as is reproduced in Table 8. You can find similar tables in the back of any high school trig book.

Let's run through a simple example of elevation-only tracking. Suppose that the distance LT is 500 feet and, for this particular hypothetical flight, the elevation angle of the model when it reaches the top of its flight is 32 degrees. The problem is solved as follows:

Table 8
Table of Tangents

Angle	Tangent	Angle	Tangent	Angle	Tangent
1	0.017	31	0.601	61	1.80
2	0.035	32	0.625	62	1.88
3	0.052	33	0.649	63	1.96
4	0.070	34	0.674	64	2.05
5	0.087	35	0.700	65	2.14
6	0.105	36	0.727	66	2.25
7	0.123	37	0.754	67	2.36
8	0.141	38	0.781	68	2.48
9	0.158	39	0.810	69	2.61
10	0.176	40	0.839	70	2.75
11	0.194	41	0.869	71	2.90
12	0.213	42	0.900	72	3.08
13	0.231	43	0.933	73	3.27
14	0.249	44	0.966	74	3.49
15	0.268	45	1.00	75	3.73
16	0.287	46	1.04	76	4.01
17	0.306	47	1.07	77	4.33
18	0.325	48	1.11	78	4.70
19	0.344	49	1.15	79	5.14
20	0.364	50	1.19	80	5.67
21	0.384	51	1.23	81	6.31
22	0.404	52	1.28	82	7.12
23	0.424	53	1.33	83	8.14
24	0.445	54	1.38	84	9.51
25	0.466	55	1.43	85	11.4
26	0.488	56	1.48	86	14.3
27	0.510	57	1.54	87	19.1
28	0.532	58	1.60	88	28.6
29	0.554	59	1.66	89	57.3
30	0.577	60	1.73	90	—

From trig table: tan 32° = 0.625
$$RL = LT \times \tan \angle\ a$$
$$= 500 \times 0.625$$
$$= 312 \text{ feet}$$

The elevation-only method assumes that the model flies absolutely vertically over the launch pad so that the ground distance LT does not change. Only a very few models will have this sort of perfect flight. Most will weathercock into the wind or otherwise deviate from a vertical flight path. If the model flies *toward* the tracking station, the distance LT becomes less and the elevation angle, for a given altitude, becomes greater; thus, the reduced altitude data will show that the model apparently flew to a higher angle and altitude than was actually the case.

This shortcoming can be eliminated to some degree by using a very long baseline LT and by locating the tracking station so that it is crosswind from the launch pad. Then, if the model weathercocks, it does not fly toward or away from the tracker and the distance LT does not change very much.

To fully account for the fact that a model may not always fly vertically, a tracking system must be able to give an achieved altitude figure no matter where the launch pad is located with respect to the tracking stations and no matter where in relation to the tracking stations the model reaches apogee. To accomplish this, a tracking device must be able to follow the model in elevation and in azimuth. The azimuth angle is a horizontal angle. A tracking device that will produce both elevation and azimuth angles is called a theodolite.

Many types of theodolites have been built and used in model rocketry. Some of them are very simple, while others have been very complex and sophisticated.

The basic parts of a tracking theodolite are shown in Figure 15-5. This assembly may be mounted on any sturdy camera tripod. It may be made from wood, metal, or plastic, depending upon the type of material and the shop facilities available. Inexpensive plastic or metal protractors available in school supply stores are perfectly adequate for use in model rocket tracking theodolites. These protractors will provide angle-measuring scales for both the elevation and azimuth axes.

Several types of optical tracking aids may be attached to this basic theodolite base. The simplest and most effective is just a long, straight piece of wood 1 inch wide by 1/2 inch thick by 18 inches to 24 inches long. A small headless brad is driven into the top side of the stick at the far end, and an ordinary screw eye is inserted in the same side at the observer's end to provide a peep sight or ring sight in which to center the top of the brad and produce an accurate line of sight. This sort of open sight arrangement with no lenses or optical magnification has been found to be excellent and highly accurate as long as the model or its smoke trail can be seen by the unaided eye.

A somewhat more elegant sighting device is a mailing tube or body tube 2 inches to 3 inches in diameter and about 18 inches long with thread cross hairs glued across both ends.

Using these simple sights, one follows the model over the top of the stick or tube during powered flight and begins to zero in on the model with the sights or cross hairs only as it begins to slow down near the top of its flight. Standard model rockets 1/2 inch in diameter and 8 inches long have been tracked for United States and international performance records to altitudes of 1,800 feet or better using this sort of simple equipment.

When it comes to mounting an optical instrument like a telescope on a tracking theodolite, you get into the realm of optics. Contrary to what you might think, don't use a high-powered telescope on a theodolite. The higher the optical power, the smaller the field of view

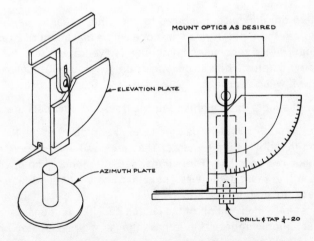

Figure 15-5: Sketch of a simple alt-azimuth tracking theodolite that can be made in most school or home workshops.

Figure 15-6: Dr. William Perendy, of Mankato, Minnesota, developed this highly accurate tracking theodolite that can be made with common hardware.

and the more difficult it becomes to keep the model in the field of view during the flight. A rifle telescope with a maximum of 2X is the most that I would recommend, although I have tracked with theodolites using surplus elbow telescopes with 10X and a 10-degree field of view—it wasn't easy!

Stick with simple open sights. You will track more models that way.

For the theodolite system using elevation and azimuth angles, two tracking theodolites are required. They are positioned at both ends of a measured base line. For national and international competition this base line must be no less than 300 meters (984 feet) long. The two tracking theodolites and the base line may be set up in any relationship to the location of the launch pad. It is not necessary and is, in fact, highly undesirable to have the launch pad located on the base line between the two stations. According to research carried out by J. Talley Guill in 1972, the optimum location of the launch pad is at an azimuth angle of 30 degrees from each station; this location provides the most accurate tracking.

The two stations must have a clear view of the launch pad and of each other. They should be roughly on the same level; in other words, don't put one of them on the top of a hill and the other one down in a valley. It is usually wise to locate the stations south of the launch pad so that neither station has to look into the sun to follow the model. This positioning also allows the model to be viewed in the full light of the sun. The tracking situation is shown diagrammatically in Figure 15-7.

With this system the theodolites are set up and leveled so that their azimuth plates are horizontal. They are then zeroed in by sighting at *each other* and adjusting their azimuth and elevation dials to read zero under these conditions. They must not be zeroed in by sighting at the launch pad. Have them look at each other to calibrate.

We now have a tracking situation with a known distance between two stations, plus, after a flight, an elevation angle and an azimuth angle from each of the two stations. This is more than enough data to compute the achieved altitude.

To understand how this is done, let's derive the equations used, referring to Figure 15-7 in the process.

Given: Distance *b*
       Angle ∠ *A*
       Angle ∠ *D*
       Angle ∠ *C*
       Angle ∠ *E*

Point *X* is an imaginary point on the ground directly beneath the model in flight. The model is located at Point *R*. We know the Distance *b*; it's the measured base line. We are trying to compute the Distance *RX*, the vertical altitude of the model.

When we compute the Distance *a* or the Distance *c*, we can theoretically locate Point *X* on the ground. Then we can compute for the two vertical triangles to find the value of *RX* in two different calculations, checking each against the other for accuracy. These two vertical right triangles (*R-X-West* and *R-X-East*) can be computed separately and the value of *RX* determined from each calculation and then averaged to produce a more accurate number.

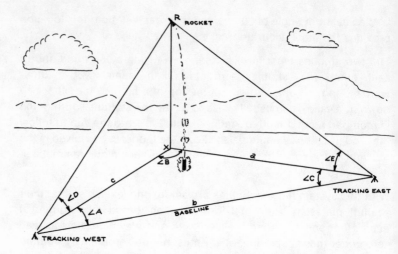

Figure 15-7: The geometry of the two-station tracking system.

The Law of Sines in trigonometry states:

$$\frac{c}{\sin \angle C} = \frac{b}{\sin \angle B} = \frac{a}{\sin \angle A}$$

Therefore:

$$c = \sin \angle C \, \frac{b}{\sin \angle B}$$

$$= \sin \angle C \, \frac{b}{\sin [180° - (A + C)]}$$

Since R is directly above X by definition, the angle R-X-$West$ is a right angle. We can therefore compute the western triangle as follows:

$$\tan \angle D = \frac{RX}{c}$$

Therefore:

$$RX = c \tan \angle D$$

Substituting for *c*, we get:

$$RX = \sin \angle C \tan \angle D \frac{b}{\sin [180° - (A + C)]}$$

The other vertical right triangle, the eastern triangle, is solved in an identical manner:

$$RX = \sin \angle A \tan \angle E \frac{b}{\sin [180° - (A + C)]}$$

The two values of *RX* thus obtained are then averaged. If either value deviates by more than 10% from the average, it can be assumed that something went wrong. It means "track lost." But if both values of *RX* come within plus or minus 10% of the average, it is considered "track closed," and the achieved altitude is taken as official.

You do not have to memorize all of this mathematical banjo music to make the system work for you. All you have to do these days is be able to read some tables, write some numbers into blanks, and do some common arithmetic. The system has been simplified, linear-programmed, computerized, and made so easy that eight-year-old model rocketeers have been able to do data reduction.

To see how easy it is, let's first work through the calculations the long way with an example:

Given: *b* = 300 meters (base line)
      Tracking East azimuth (∠ *C*) = 23°
      Tracking East elevation (∠ *E*) = 36°
      Tracking West azimuth (∠ *A*) = 45°
      Tracking West elevation (∠ *D*) = 53°

For one triangle, we figure:

$$RX = \sin \angle C \tan \angle D \frac{b}{\sin [180° - (A + C)]}$$

$$= \sin 23° \tan 53° \frac{300}{\sin [180° - (45° + 23°)]}$$

$$= 0.391 \times 1.33 \times \frac{300}{\sin 68°}$$

$$= 0.391 \times 1.33 \times 324$$
$$= 168 \text{ meters (551 feet)}$$

Y.M.C.A. SPACE PIONEERS

NAME _____ DATE _____

NAR # _____

MODEL _____ MOTOR TYPE _____

LAUNCH AREA # _____

SAFETY CHECK BY _____

DATA REDUCTION BY TRIGONOMETRY

Tracking #1 Azimuth _____° sin _____ ① Elevation _____° tan _____②

Tracking #2 Azimuth _____° sin _____ ③ Elevation _____° tan _____④

Add azimuths for∠B _____Table Value V for∠B _____⑤

Multiply _____ ⑤ ×_____ ② ×_____ ③ = _____

Multiply _____ ⑤ ×_____ ④ ×_____ ① = _____

ADD: _____

This is Average Altitude──→Divide by 2: _____

10% RULE CALCULATION:

Average Altitude_____

ADD 10%

of Average Altitude_____

Result_____

If result is LESS than highest of the two computed altitudes, TRACK LOST.

_____ OK _____ LOST

Average Altitude_____

SUBTRACT 10%

of Average Altitude_____

Result: _____

If result is MORE THAN lowest of the two computed altitudes, TRACK LOST.

_____ OK _____ LOST

Figure 15-8: Flight data sheet used in data reduction.

Solving the other triangle by the same method gives us RX = 167 meters (548 feet).

The average of these two altitudes is 168 plus 167 divided by 2; this gives the result of 167.5 meters. In applying the 10% Rule, the average altitude is rounded off, and the rule here is: Keep it even. If the average altitude ends in an even number before the decimal point, the resulting decimal is dropped and the number is "kept." If the number is odd, it is rounded off up to the next even-numbered meter. In this case the average altitude would be 168 meters. Both computed altitudes must fall within 10% of this figure. In our case 10% is 16.8 meters. Since both altitudes are within 16.8 meters of the average, the track is considered good in this example.

This system was originally worked out in 1958 by Art Ballah and Grant Gray. It was refined into a very rapid system in 1960 by John S. Roe. He devised a table giving both sine and tangent values, plus a column for the value of 300/sin∠ B. See Table 9. This method is very

Table 9
Altitude Calculation Table

Use this table for computing the altitude of models using elevation and azimuth angles from two tracking stations on a base line of 300 meters, or 984.24 feet. V = 300/sin B.

Angle	Sin	Tan	V
1	0.0174	0.0174	17189.6
2	0.0349	0.0349	8596.11
3	0.0523	0.0524	5732.2
4	0.0698	0.0693	4300.7
5	0.0872	0.0875	3442.1
6	0.1045	0.1051	2870.0
7	0.1219	0.1228	2461.6
8	0.1392	0.1405	2155.6
9	0.1564	0.1584	1917.7
10	0.1736	0.1763	1727.6
11	0.1908	0.1944	1572.3
12	0.2079	0.2126	1442.9
13	0.2249	0.2309	1333.6
14	0.2419	0.2493	1240.1
15	0.2588	0.2679	1159.1
16	0.2756	0.2867	1088.4
17	0.2924	0.3057	1026.1
18	0.3090	0.3249	970.8
19	0.3256	0.3443	921.47
20	0.3420	0.3640	877.14
21	0.3584	0.3839	837.13
22	0.3746	0.4040	800.84
23	0.3907	0.4245	767.79
24	0.4067	0.4452	737.58
25	0.4226	0.4663	709.86
26	0.4384	0.4877	684.35
27	0.4540	0.5095	660.81
28	0.4695	0.5317	639.02
29	0.4848	0.5543	618.80
30	0.5000	0.5774	600.00
31	0.5150	0.6009	582.48
32	0.5299	0.6249	566.12
33	0.5446	0.6494	550.82
34	0.5592	0.6745	536.49
35	0.5736	0.7002	523.03

Angle	Sin	Tan	V
36	0.5878	0.7265	510.39
37	0.6018	0.7535	498.49
38	0.6157	0.7813	487.28
39	0.6293	0.8098	476.70
40	0.6428	0.8391	466.72
41	0.6561	0.8693	457.28
42	0.6691	0.9004	448.34
43	0.6870	0.9325	439.88
44	0.6947	0.9657	431.87
45	0.7071	1.0000	424.26
46	0.7193	1.0355	417.05
47	0.7313	1.0723	410.20
48	0.7431	1.1106	403.69
49	0.7547	1.1504	397.50
50	0.7660	1.1918	391.62
51	0.7715	1.2349	386.03
52	0.7880	1.2799	380.71
53	0.7986	1.3270	375.64
54	0.8090	1.3764	370.82
55	0.8192	1.4281	366.23
56	0.8290	1.4826	361.87
57	0.8387	1.5399	357.71
58	0.8480	1.6003	353.75
59	0.8572	1.6643	349.99
60	0.8660	1.7321	346.41
61	0.8746	1.8040	343.01
62	0.8829	1.8807	339.77
63	0.8910	1.9626	336.70
64	0.8988	2.0503	333.78
65	0.9063	2.1445	331.01
66	0.9135	2.2460	328.39
67	0.9205	2.3558	325.91
68	0.9272	2.4751	323.56
69	0.9336	2.6051	321.34
70	0.9397	2.7475	319.25
71	0.9455	2.9042	317.29
72	0.9511	3.0777	315.44
73	0.9563	3.2709	313.71
74	0.9613	3.4874	312.09
75	0.9659	3.7320	310.58
76	0.9703	4.0108	309.18
77	0.9744	4.3315	307.89

Angle	Sin	Tan	V
78	0.9781	4.7046	306.70
79	0.9816	5.1445	305.62
80	0.9848	5.6713	304.63
81	0.9877	6.3138	303.74
82	0.9903	7.1154	302.95
83	0.9925	8.1443	302.25
84	0.9945	9.5144	301.65
85	0.9962	11.4300	301.15
86	0.9976	14.301	300.73
87	0.9986	19.081	300.41
88	0.9994	28.656	300.18
89	0.9998	57.29	300.04
90	1.0000	∞	300.00

fast and requires only a slide rule or pocket electronic calculator. You can also do the calculations the long way on a piece of paper. This is still quite fast and is widely used because of its simplicity.

Naturally, the next thing to do was to put the whole situation into one of the modern high-speed digital computers and instruct it to solve for every possible combination of azimuth and elevation angles. This programming was first done by J. Talley Guill at Rice University, and a whole line of computerized precomputed altitude tables followed. These tables were used quite extensively at national meets for many years.

Another tracking system has been developed by J. Talley Guill, who felt it should be possible to have a tracking system using several elevation-only trackers like the one described first in this chapter. These trackers are simpler to make and easier to operate than the ones that measure both elevation and azimuth angles. Guill worked out his new system in the summer of 1972. Three elevation-only trackers are required, and they are located on a single base line, as shown in Figure 15-9, with one tracker at each end and the third precisely in the center of the base line. The system was checked out and directly compared with an operating, calibrated two-station azimuth-elevation tracking system. The three-station elevation-only system matched the other in all respects and produced the same altitudes. The Guill System is detailed in Appendix V. It has been

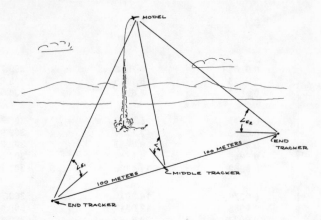

Figure 15-9: The geometry of the three-station elevation-only tracking system. See Appendix V.

tested, and it works. But it has not been extensively used yet.

When using tracking stations on a model rocket range, you must have some means of communicating between them and with the range control point. Sometimes no communication system other than flags is used. There is a hand-held flag at range control and one at each tracking station. When a model is ready to go, range control lifts its flag. The trackers see this and raise their flags, indicating they are ready. When the range control flag drops, the model is launched. The trackers then either write down their angles or send them back to range control via runner. This simple flag communications system was first used by the Czechoslovaks at the First International Model Rocket Competition in Dubnica, Czechoslovakia, in May 1966.

More sophisticated range communications systems were developed in the United States from 1958 onward. Basically, these were hard-wire or land-line telephone systems. At first, surplus Army field telephone wire was used; this stuff is cheap and tough, designed to be run over by Main Battle Tanks. It can lie out in the field for years without damage. I obtained several miles of it from the 1960 Boy Scouts Jamboree in Colorado. Believe it or not, five different model rocket ranges were established with that wire! I must have rolled out and rolled in several thousand miles of Army field telephone wire over the years. Sometimes I've been able to talk the local telephone company into donating telephone handsets and headsets to use on the range. Sometimes I had to buy them. Telephone systems are great! They are reliable and permit a tracking station operator to talk to the launch area and call "Hold!" if there is trouble, even during a countdown.

But modern electronics and integrated circuits may have finally caught up with the model rocket field telephone system. Citizen's Band walkie-talkie radiotelephones have now become so cheap and easy to get that they have taken over from the old field telephones. They can provide good communications with the trackers and range control. All three units must be on the same frequency, and they must have enough power to punch through all the yak-yak on the Citizen's Band. The cheapest units will not do the job. You will need a couple of watts of power to make this radio system reliable. Radiotelephones offer much easier range operations with a lot less equipment, even though their cost may be about the same or slightly more.

Now, with trackers and communications, you almost have a full-fledged model rocket range!

16

Model Rocket Ranges

Although you may begin flying model rockets in an open field with one or two other people, you will soon be joined by others who wish to fly also. You can have a lot more fun and learn more when model rocketeers get together to fly. However, if many models are flown and if they are to be tracked for altitude, some additional equipment will be required and you will have to have some form of organization to maintain safe operations and to prevent confusion.

Many people upon first visiting a model rocket range are impressed by its organization and safety control. These range characteristics are no accident; they did not evolve haphazardly. The first time model rocketeers got together to fly, they immediately recognized the need for organization and safety control. The reason for this awareness was that we early model rocketeers were all professional rocketeers from White Sands Proving Grounds in New Mexico, and we had all been thoroughly indoctrinated with the safety procedures used in flying real rockets. Very few model rocketeers realize the tremendous debt they owe to two men who pioneered the flight safety practices of full-scale rocketry and astronautics, and whose policies were adopted into model rocketry. These two men are Herbert L. Karsch and Nathan Wagner, the flight safety engineers at White Sands in those days. Much of what you are going to read here was adopted straight from White Sands, where all of us got our early training in rocketry because it was the only place to do it then. It is largely because of Karsch and Wagner that model rocketry, since its beginnings in 1957, has conclusively proven to be safer than model airplane flying and safer even than bicycle riding.

Figure 16-1: A model rocket range is a place of organized activity where modelers can fly their rockets in safety and gain more information about rocket performances.

The number one safety rule taken over from professional rocketry is: The Range Safety Officer's word is as the word of God, and nobody can override a safety decision made by the Range Safety Officer. If he says, "No," that's it. Either do what he says or go someplace else.

At White Sands, and later at Cape Canaveral, Vandenberg, Point Mugu, Eglin, and Wallops, the Range Safety Officer is supreme. He or his trained crew checks every rocket vehicle before flight. He alone determines whether or not it will be safe to fly. He gives the final safety clearance before the launch. He even destroys the vehicle in the air if it becomes unsafe.

The basic range philosophy is: Don't take chances. It isn't worth it. That is one of the reasons rocketry is still "the world's safest business."

Those of us who started model rocketry profited from the experiences of the professional range safety people and adopted as many of their practices as practical. The only thing we don't do in model rocketry is destroy the model in midair if it becomes unsafe.

While adopting professional range safety procedures for model rocket ranges, we also borrowed many other aspects of life-size rocket flight testing. Range organization was one such aspect. Although model rocketry is a very individualistic hobby where a person can use his creativity to the utmost within the bounds imposed by reality in the design and construction of his model, it is also a team effort on the model rocket range. Everyone works together on the range.

A model rocket range is a little Cape Canaveral. It is run the way the Cape is. If anyone who had a rocket to fly were allowed to set up his launch pad anywhere at the Cape and fly whenever he wanted to, there would be chaos. The same holds true for a model rocket range. Not only is it fun to work together in a professional manner, but operating in this way helps to tie our hobby closer to the real thing.

The first requirement for a model rocket range is a large open space of land. For models flying up to 1,000 feet, an ordinary school football field is adequate if it isn't hemmed in by too many rocket-eating trees. If you are going to fly large models or do a lot of glide-recovery work or payload flying, you must have a larger field. Model rocketeers face the same problem as model aviators: Where can we find a place large enough for flying? In the western and midwestern parts of the United States away from urban and suburban areas, model rocketeers usually have no trouble. Somebody always has a friend or relative on a nearby farm. But this is not true if you are part of the 90% of today's American people who live in urban and suburban areas.

There are no pat answers to the problem of finding a place to fly. But you should always have permission of the owner of the land. In my own experience it has often been easier to get permission to use private land such as a farm field. However, many model rocket clubs that are sponsored by civic and youth groups have received the backing of local park and recreation departments. This has resulted in permission to use school and park grounds.

For you young model rocketeers the most important single requirement for model rocket flying is an adult who is interested and willing to act for you. An adult will know how to approach local authorities

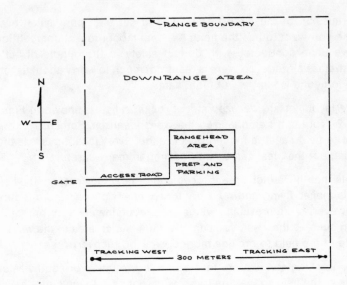

Figure 16-2: Layout of an idealized model rocket range.

and can speak as a tax-paying, voting citizen. Believe me, this helps! Your adult spokesman can also be your Range Safety Officer, club advisor, or club sponsor. More about that in the next chapter.

Any flying field that you select should be reasonably free of trees, for reasons already discussed. It should also be clear of high-voltage electric transmission lines; any model stuck in an electric line should be left there because you may be killed trying to remove it.

Your flying field should also be away from heavily traveled roads and high-speed highways, parkways, turnpikes, and freeways. A model rocket will not damage an automobile, but if a parachuting model or glider wafts down in front of a driver speeding along in heavy traffic, it may severely startle him and cause an accident.

By all means, stay away from airports, airport runway approach lanes, and areas frequented by low-flying aircraft. A pilot who is landing or taking off has plenty to do without worrying about whether or not your model rocket may hit his plane. Although it is practically impossible to hit an airplane in flight (one chance in 50 billion flights), your model could startle or distract a pilot at a critical moment.

If your small local airport is used mostly by private aircraft and is the only place around where you can fly, fly there only with the permission of the airport officials, and get their permission every time you want to use the field. National model rocket competitions have been conducted with perfect safety in the middle of active first-class airports, so there is really not much to worry about as long as someone is alert about flight safety.

The layout of an *idealized* model rocket range is shown in Figure 16-2. Note that I emphasized the word *idealized*. Someday I would like to fly on such a model rocket range. Every location is less than ideal, but the ideal can be approached in many cases.

Note that the launch area is located at or near the center of the field. This makes flying and recovery independent of the wind direction. However, if the prevailing winds are predominately from one direction such as the west, you can offset the launch area to the western side of the field to provide more downwind recovery area.

The tracking stations are located on the southern or southwestern side of the field so that the trackers do not have to look into the sun when following the models in flight. Note that the trackers have a clear and unobstructed view of the launch area at all times.

The launch area normally faces north or east so that the Range Safety Officer, Range Control Officer, and model rocketeers have the sun at their backs when flying model rockets. You may not think this point is very important until you have been on the range for a whole afternoon and suffered sunburned eyeballs from having looked continually into the sun! A couple of national meets were set up with the launch area facing the sun, resulting in five rather miserable days.

There should be only one entrance to the flying field, or model rocket range. A sign should be posted at the entrance to let people know what is going on and to advise them to stay alert if they enter the field. The usual sign says:

PODUNK MODEL ROCKET CLUB
FLYING FIELD
Model Rockets in Flight
PROCEED WITH CAUTION

Model rocket range signs should *never* say, "Danger! Look Out for Falling Rockets!" or other such things, because it isn't so.

Cars should be parked so they don't block the line of sight from the trackers to the launch area.

The launch area is usually called the rangehead area. Figure 16-3 shows the layout of a typical modern rangehead area. This system was initially worked out in 1965 by the New Canaan (Connecticut) YMCA Space Pioneers Section of the NAR. It replaced an older system that used launching racks and was less flexible. The first Sunday afternoon that we Pioneers tried it out, Dick Ploss of the New Canaan YMCA humorously tagged it "Misfire Alley." The name stuck, and the layout is known by that name today around the world!

The rangehead area is separated from the spectator and prep areas by a simple rope barricade. There is an open "gate" in the center. The barricade keeps spectators out of the launching part of the rangehead area. Everyone who enters or leaves the launching section should use the gate and not jump the barricade.

The barricade can be very simple, almost symbolic, in nature. Often a local gasoline station will give you some of its used or extra bannered ropes, the kind with plastic pennants hanging from them. These are colorful and attractive. They are made into a barricade by simple wooden dowels pounded into the ground to support them. There should be a dowel post every 10 feet or so, and you will probably need at least two 50-foot lengths of banner line.

Fifteen yards away from the barricade are the launch areas. Each is marked with a large sign giving the launch area number. These can be plywood signs affixed to wooden posts that are temporarily pounded into the ground. There should be 15 feet to 20 feet (5 paces) between the launch areas for safety.

Each launch area should have a canvas tarp at least 5 feet square on which a launcher can be set up. The tarp keeps your clothing clean while you are down on your knees working around the launch pad hooking up your model. It also prevents grass fires if the field is dry and grassy. On March 31, 1967, I was on hand when our club accidentally burned off an 11-acre grass field. All safety precautions were in effect, but a glowing piece of igniter landed in the grass next to the pad. A 15 mile-per-hour wind was blowing that day, and the blaze was out of control by the time the Safety Officer saw it a few seconds after launch. This one mishap was enough for everyone in the club. We purchased tarps for each launch site. And we *never* had another grass fire in the remaining six years that I was with the club. Those tarps were worth every penny!

Two portable folding tables are set up next to the barricade gate, as shown in Figure 16-3. At one of these tables the Range Safety Officer

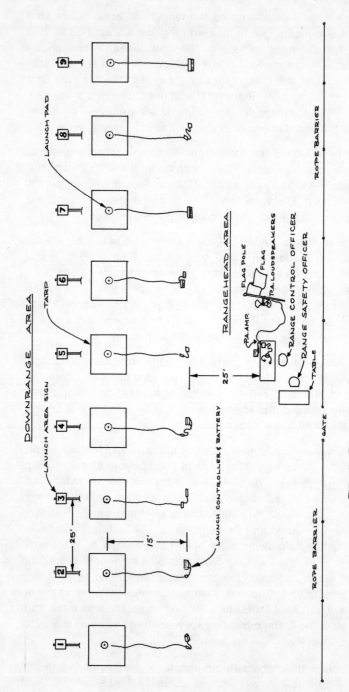

Figure 16-3: Diagram of the rangehead area of Misfire Alley.

presides. Every model that is to be flown is taken into the rangehead area and presented to the Range Safety Officer (RSO) at his table. During contests this table is also used as the check-out table for working up the flight cards. There may also be a weighing scale on the table if there is a weight limit on the models, such as in competitions.

The other table provides a place for the Range Control Officer (RCO) to work. He sits there in command of the entire range. The RCO has a *loud* public address system and microphone. He also has communication with the tracking stations by walkie-talkie radios or land-line telephones.

There are several types of battery-operated public address (PA) systems on the market today. But it is also very easy to build your own. A schematic for a fully transistorized PA system is shown in Figure 16-4. It can be built with parts that can be purchased at most radio stores or ordered from Allied Radio or Radio Shack. It can operate from any 12-volt auto or motorcycle battery. I built a system exactly like this in 1966. It was still working without any maintenance in 1973.

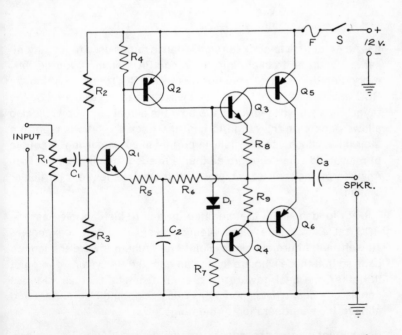

Figure 16-4a: Schematic diagram of a transistorized, battery-operated 10-watt PA amplifier.

C1:	10 mfd. 15 vdc	D1:	1N2069
C2:	100 mfd. 15 vdc	R1:	500K volume control
C3:	1000 mfd. 15 vdc	R2:	39K, $^1/_2$ w.
J1:	Input jack	R3:	39K, $^1/_2$ w.
J2:	Loudspeaker jack	R4:	1000 ohms, $^1/_2$ w.
J3:	Battery jack	R5:	1000 ohms, $^1/_2$ w.
Q1:	2N1694	R6:	1000 ohms, $^1/_2$ w.
Q2:	2N1305	R7:	680 ohms, $^1/_2$ w.
Q3:	2N1304	R8:	47 ohms, $^1/_2$ w.
Q4:	2N1305	R9:	47 ohms, $^1/_2$ w.
Q5:	2N554	F:	2 amp. fuse and fuse holder
Q6:	2N554	S:	SPST power switch

Notes:
1. 2N554 power transistors must be mounted on heat sinks.
2. Amplifier will operate on battery voltages from 3 v. to 15 v.
3. Reversing battery polarity may destroy transistors.
4. Amplifier will operate with loudspeaker impedance from 4 to 16 ohms. Operation without an attached loudspeaker may destroy the amplifier.
5. Input of amplifier is hooked in parallel (bridged) across communications lines. If DC phone system is used, another 10 mfd. 15 vdc capacitor should be placed ahead of R1.

Figure 16-4b: Parts list for the transistorized PA amplifier.

It is no great trick to hook the walkie-talkie radios used for communications with the trackers into the range PA system. If you do this, everything that is said to the trackers by the RCO and everything the RCO broadcasts on the PA system will be heard by everyone on the range. This allows the trackers to hear the actual countdown. It also allows people on the range to hear the tracking stations reporting their data, which can be vitally important in contest flying. A couple of relays and some wire are all that is needed to make the hookup. Any sharp electronicist can work out the details for your particular system.

The PA loudspeakers are mounted on a portable flagpole near the RCO table. Note the plural—loudspeakers. Get two inexpensive 10-watt metal horn speakers. Point one of them toward the launch pads so that those who are flying can hear the PA system, and point the other one back toward the prep and spectator areas so those people can hear what is going on. This positioning is important and adds to the safe operation of the range.

Don't point one of the speakers toward the RCO and his microphone; this will cause the system to squeal.

It is vitally important to have a *loud* PA system on the range and to have good communications. They help the RSO and the RCO maintain range discipline.

A good range should have a flagpole. It can be portable. A typical portable flagpole made from pipe and other common hardware store items is shown in Figure 16-5. It is a good idea to fly a flag, either the national colors or your club insignia. A flag is a very good wind indicator, and it draws attention to the fact that something is going on in your flying field. Also, after a long trek back from recovering a model, it shows you where the rangehead area is. The flag idea was adopted from White Sands, which adopted it from rifle and artillery ranges where a red flag is always flown when the range is in operation.

The equipment discussed so far will cost about $100 to $150, depending upon how much can be located in the junk piles of

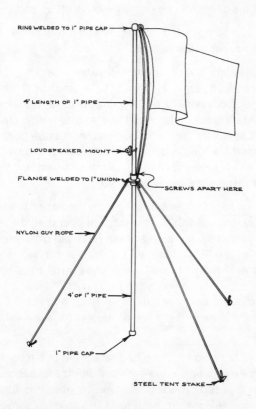

Figure 16-5: Sketch of a portable flagpole.

members' workshops and how much can be built by handy members. It is not difficult to design and build all of the equipment so that it breaks down into small portable modules that can be packed away for transportation and storage in specially made wooden boxes. You should strive to make it compact to the point where it can be carried in the back of a single station wagon, and no box should be larger and heavier than can be easily lifted by two people.

By all means, build your range equipment rugged. It will take a beating. If you build it with ruggedness in mind, it will last for nearly a decade of hard use with only routine maintenance.

How does a range such as this operate?

The first order of business when everyone arrives at the start of the day's flying activities is to set up the range. Actually, as few as six people can do the job. However, the more people, the faster it goes together. Nobody should be permitted to prep his models or set up anything of his own until the entire range is set up and checked out.

The range can be operated by one person if manpower is short. At the most, it requires four people.

The Range Safety Officer performs safety-checking and control duties as described earlier. He should be an adult or at least eighteen years of age. Occasionally he is going to have to make a difficult decision that somebody else won't like, and he is going to have to be able to stick by it, often under considerable pressure from adults and parents. Therefore, long and hard experience has shown that only an adult or older rocketeer should be permitted to assume the position, duties, and responsibilities of the RSO.

The Range Control Officer can be any club member. While the job of the RSO is safety, the RCO is in charge of operating the range. If the range isn't very busy, the duties of the RCO can often be carried out by the RSO. The RCO talks to the trackers, gives countdowns, selects the next model to be launched, records data, and keeps a running commentary going on the PA system. It takes a person with a steady hand and cool head to be an RCO, and many people have learned how to keep calm in a hassle because of their RCO experience.

Two trackers are also required to operate the two tracking theodolites if tracking is being done—three if the three-station elevation-only system is used. There is no need to man the stations if no tracking is to be done; the range then operates without altitude tracking capabilities as a fun, sport, or duration range.

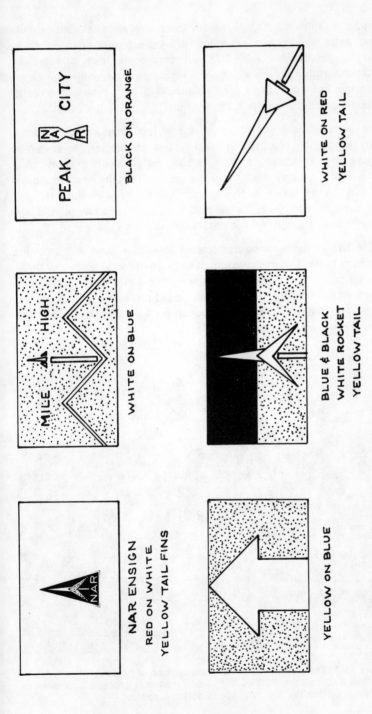

Figure 16-6: Some typical NAR club and range flags.

Once the range is set up and ready to go, everybody should sign up on a range manpower sheet at the RSO table; people are required to carry out the duties of RSO, RCO, and trackers for thirty-minute time periods throughout the afternoon. A typical range manpower sheet is shown in Figure 16-8. Everyone should take his turn on the range at some point during the flight session.

Everyone who is going to fly picks a launch point and sets up his own launch pad, electrical ignition system, and battery, as shown in Figure 16-3. This personal responsibility system in which everyone is in charge of his own equipment has proven itself the best procedure in club flying. Since each member builds, uses, maintains, and improves his own GSE, it's his own fault if it doesn't work. And it is usually working right by the next flight session.

Often two or more model rocketeers share the same launch point because there is plenty of room at each point to set up two or three launch pads. There are seldom any conflicts in doing this because rarely do two modelers want to launch at the same time out of the same launch area. When it does occur, no problems arise if the RCO does his job and assigns priorities.

Figure 16-7: The rangehead area is always one of purposeful activity with the RSO and the RCO on duty while modelers process their models and timers clock the flight durations during a contest.

There are two cardinal rules in the rangehead area:
1. Never cut across the launch alleys. You could trip over somebody else's ignition leads, and this makes him very angry.
2. Always approach the launch points at right angles to the line of launch pads. Keep your head up and listen for any countdown. The RCO will be watching to ensure that you do not walk into a launch area that is into its terminal countdown, but you should also keep your eyes open.

Occasionally for a large model, the RSO will order adjacent launch points to be vacated during the final countdown.

When you have prepped your model in the prep area and are ready to fly, you bring the model into the rangehead area through the gate in the barrier and present it to the RSO for its safety check. This is *always* done. The RSO, upon clearing your model for flight, permits you to proceed to your launch point. Place your model on your own launch pad, hook it up, step back to the firing point, and raise your hand. The RCO keeps a continual watch on the launching line for people with their hands up, signaling that they are ready to launch. Normally, he will work the launch points in rotation, starting with No.

YMCA Space Pioneers	RANGE MANPOWER						Date: _____ Contest: _____		
Time	Range Safety	Range Control	Track N	Track S	Return	Timers	Fire Guard	D	R
1:30-2:00									
2:00-2:30									
2:30-3:00									
3:00-3:30									
3:30-4:00									
4:00-4:30									
4:30-5:00									

Figure 16-8: The Range Manpower Sheet developed for use on the Misfire Alley range.

1 and proceeding to the other end, then coming back to pick up at No. 1 again. This is the fair way to operate where there are lots of launches. Sometimes you must wait a few minutes until the RCO gets around to you.

When your turn comes, the RCO gets safety clearance from the RSO. You may then insert your safety key into your launch controller. The RCO goes into the terminal countdown over the PA system. When he reaches zero, you push your own launch button and launch your own model. It is then up to you to recover and return your own model.

The Misfire Alley system has been working since it was first tried out in 1965. It has proven itself to be completely safe in operation. There is no need to get expensive club launching equipment or a centrally switched electrical ignition system; these things bog down the range operations and can actually be more hazardous than the individual control for which Misfire Alley was designed. If somebody launches without a safety clearance and a countdown, the RSO declares him out of action until the modeler finds the trouble, fixes it, and checks the fix with the RSO. Yes, there are occasional accidental launches. But, because of the PA system, the RCO, and the RSO, I have never seen one of these accidental launches become hazardous in any way. Such a launch causes so much embarrassment to the model rocketeer that he quickly gets his equipment into safe shape.

If you were to hear a recording of the flight operations of a range, it might sound something like this:

RCO: "Okay, the next bird to go is from Launch Area Number Six. It is a single-staged bird with a Type B motor and painted fluorescent orange overall. Stations report."

Tracking East: "East go."

Tracking West: "West go."

RSO: "Safety go."

RCO: "Range is go! Safety is go! Arm the panel! Time is running at T-minus five . . . four . . . three . . . two . . . one . . . *start!*"

(The term "start" is a universal term used internationally to indicate the instant of ignition. It means nearly the same thing in all languages. It is preferable to "zero" or "launch." The word "fire" must *never* be used on a model rocket range unless there is a grass fire—and don't forget this!)

RCO: "Model is coming up on peak. *Mark!* (The word "mark" indicates the moment of maximum altitude as seen from the rangehead area; both trackers should stop tracking at "mark" and lock their theodolites.) Recovery system is deployed, and the descent looks good. Okay, trackers, your angles, please."

East: "This is Tracking East. Azimuth, three-two degrees."

RCO: "Roger, East. Azimuth, three-two degrees." (The angles are always reported this way, and the RCO repeats them to make sure that he has heard them correctly.)

East: "Elevation, two-seven degrees."

RCO: "Roger. Elevation, two-seven degrees."

West: "This is Tracking West. Azimuth, three-five degrees."

RCO: "Roger, West. Azimuth, three-five degrees."

West: "Elevation, three-zero degrees."

RCO: "Say again elevation."

West: "Elevation, three-zero degrees."

RCO: "Roger. Elevation, three-seven degrees."

West: "Negative, Range Control. Elevation is three-*zero* degrees."

RCO: "Sorry. Elevation, three-zero degrees. The next bird to go . . ."

And so it goes, all afternoon. On a well-run range as many as fifty models can be launched every hour from a twelve-point rangehead.

If during the preflight or terminal countdown something goes wrong, *anybody* on the range may yell, "HOLD!" When the RSO and the RCO hear this, they freeze the countdown right there, recycle the countdown back to the range safety clearance point, and find out what the reason for the hold is. This is an additional safety measure since it makes everyone on the range a "deputy range safety officer" who can call a halt if he sees something wrong. Naturally, this procedure could get out of hand without good range discipline. Therefore, the RSO must *never* permit horseplay or false alarms on a model rocket range. It is his responsibility to maintain order and discipline on the range.

Some flights are not successful. But, because of good range safety discipline, I have never seen any serious hazard on such a model rocket range. You must always keep in mind, however, that you are

Figure 16-9: Finally, don't forget that you have to recover your own model. Often a loyal and trustworthy recovery crew can be a model rocketeer's best friend.

in a big open area with objects that are flying freely through the air. Therefore, I pass on to you the following tips from my own experience in watching well over 100,000 model rocket flights:

1. If you do not have range duties that require your attention, stop what you are doing during the terminal countdown and watch the model. Keep your eyes on it until you know where it is going to land and that the flight is going well.

2. Get on your feet and stay on your feet. Don't lie down and take a nap in the warm sun. You may have to move quickly to get out of the way of an errant model. It doesn't often happen, but when it does, be prepared.

3. If a model gets into trouble in the air, do not panic and start to run. Stand still and keep your eyes on it. If it comes your direction, you will have to move less than twelve inches to one side to get out of its way.

4. Do not engage in horseplay or practical jokes on the range at any time, and don't let others do it.

5. Don't crowd other people. During a heads-up flight, stand at least an arm's length away from everyone else.

6. If a model falls to the ground before its ejection charge has gone off, stand clear of it until this occurs. Then you may pick it up.

7. When recovering a model, do not run up to it. You may trip and step on it. Don't recover somebody else's model unless he asks you to or you find it 'way out in the boondocks. If you are recovering your own model, it's okay to snag it out of the air before it touches the ground, but don't do it with somebody else's model unless he asks you to in advance. Let a model drop to the ground; then pick it up carefully.

8. Keep a clean model rocket range. Provide trash cans, and be tough on people who don't use them.

9. Help the RCO and the RSO keep spectators under control. Don't let them wander into the rangehead area or into the downrange recovery area. You and your group are responsible for the safety of everyone on the field. Your club will be a better one because of good range discipline, and the spectators will respect your group because of it.

10. Keep dogs and small children under control at all times.

11. Don't even try to fly on days with high winds and bone-dry grass.

12. Use common sense, keep your cool, and have fun.

I have been on many model rocket ranges all over the United States and Europe. It is always a pleasure to be on a well-run range and very disturbing to be on a poorly run range. It doesn't take much to get a poorly operating range in the groove again, and a smooth-running model rocket range can be a source of great pride.

Clubs and Contests

There is an old saying among model rocketeers: When two model rocketeers get together, they form a club.

Actually, it takes more than two people to form a going model rocket club, and a club has a lot of advantages. If you do not belong to one, you should seriously consider forming one or joining one.

It's easier to join a club already in existence. But, if there isn't one, it is not difficult to form your own club. It may come into being spontaneously when several people meet on the same field to fly together.

How do you get a great model rocket club going, how do you keep it going, and how do you make it a worthwhile activity for its members? Let's suppose that you are a lone individual who has become interested in model rockets. You've done some flying on your own, observing the proper safety rules. You'd like to learn more about the hobby, discover what other people are doing, and encourage others to do it the right way.

If you are a young person, the first thing you must do is find an adult who will act as supervisor or advisor. As I pointed out in the preceding chapter, there is much that he can do that you can't. As you'll see, there is even more to it than that. This adult might be a parent, science teacher, Scout leader, or any adult who is interested in young people and/or model rocketry.

You should also find a sponsoring organization. A sponsor can help you with a meeting place, flying site, and finances. There are many

316

organizations today that have model rocketry programs within their structure or that warmly encourage model rocket clubs. The Civil Air Patrol has an active model rocket program, and many Scout and Explorer groups are deeply involved in model rocketry. The 4-H Clubs also have a model rocket program. But there are other prime candidates for sponsors of independent model rocket clubs. These include, but are not limited to: Lions, Kiwanis, Rotary, your local fire department, local police department, and the YMCA and YMHA.

It will help in getting organized and obtaining sponsorship if you, your advisor, and your club members are all members of the National Association of Rocketry. The NAR is the nationwide organization of model rocketeers in the United States. You can write to the NAR at the headquarters address listed in Appendix I. It will send you membership information and blanks as well as information on how to get an NAR charter for your club.

I am proud to have been one of the founders of the NAR in 1957. It has grown from an idea to an organization receiving national and international respect and admiration. Today it has thousands of members and chartered Sections all over the United States. It is affiliated with the National Aeronautic Association, America's aerospace sport club. Through this affiliation it is linked directly to the Federation Aeronautique Internationale in Paris, France. National aerospace clubs from sixty-one countries belong to the FAI, and each has its national counterpart to the NAR with whom the NAR is in contact.

The NAR does many jobs. It is a nonprofit organization, which means that it is a labor of love for the people who run it.

The most important offering of the NAR for club purposes is liability insurance to protect you and your club. It will cover personal injury and property damage caused by a model if there is an accident. Having insurance does not mean that model rocketry is dangerous; it is simply protection, plus a recognition that model rocketeers are so safe that they are insured sight unseen by a blanket liability policy issued by a major domestic insurance company.

Having this liability insurance is a tremendous help in obtaining the support of sponsoring organizations or permission to use meeting places and flying fields. For a few dollars extra, the insurance can be extended to cover the sponsoring organization and the owner of the flying site. Believe me, this NAR insurance is strong help when you are looking for support! Full details are available from the NAR.

Figure 17-1: Regular club meetings bring together model rocketeers with common interests to discuss technical problems and plan joint activities.

Figure 17-2: The National Association of Rocketry offers the model rocketeer many benefits for his membership, and many clubs are chartered Sections of the NAR. Here are some of the NAR publications. Jacket patches came from various Sections, NARAMs, and international meets.

The NAR also charters local Sections, or clubs. It is not easy to get an NAR Section Charter. Stiff requirements must be met by the club. It must have at least ten members, one of whom must be an adult member of the NAR. The club must petition the NAR for a charter, listing the names and NAR membership license numbers of the club members. It must submit a set of operating rules called bylaws that must be checked and approved by the NAR. The activities of the group must be reported to the NAR at regular intervals.

The NAR wants its Sections to be active, operating groups of serious model rocketeers located in a specific area. NAR Sections are not flash-in-the-pan neighborhood rocket clubs that have sprung up overnight and are likely to fold as quickly as they were formed. Sections must have organization, adult advisors, proper direction, and the ability to last.

Bylaws should be adopted at your earliest meetings. They are the operating rules of the group, a document that the club members turn to for guidance. A sample set of recommended NAR Section Bylaws is presented in Appendix VI. Some club bylaws are simpler than this, some more complex. But any set should at least contain the provisions given in this example. Note that the bylaws do not contain any special rules for the operation of a model rocket range or for the construction and operation of model rockets by members. The basic NAR rules and standards cover these operational aspects. Special club standards may be adopted, but they should be promulgated by the club advisor, a board of directors, or an operations committee.

Note also that flight sessions are not considered to be meetings of the club. It is impossible to conduct a business meeting on a model rocket range; everyone is too busy with his models or range duties. Business meetings of the club are very important and should be held in a place where people can be comfortable and discussions can be held without distraction.

Although some financial support may come from the sponsoring organization from time to time, a club should be financially solvent. It should have its own membership dues to cover the costs of mailing meeting notices and other general club operating expenses. Most clubs charge nominal dues of a couple of dollars per year. These dues should be over and above dues to other organizations such as the NAR.

One of the most important aspects of good club operations is good communications between officers and members. There should be a

Figure 17-3: A regular club training course was developed around earlier editions of this book. For eight years the author taught most of this training course, which helped develop an outstanding model rocket club.

means for rapidly getting information to everyone. This may be done by a telephone committee where each committee member calls five or ten club members when he gets the word from the telephone committee chairman. Dates for meetings and flight sessions should be set up well in advance and should be regularly scheduled. If there is a change in plans, such as weather cancellation of a flight session, the telephone committee swings into action.

A club newsletter should not be a substitute for face-to-face communications at a club meeting. A newsletter is more a means of letting *other people* know what your club is doing! It should be sent to all members as well as to local newspapers, public officials, and officers of the sponsoring organization.

Business meetings should be short and to the point without getting tangled up in Roberts' Rules of Order. Elect a president or chairman who can keep things moving and who does not like to listen to himself ramble on. Get in touch with NASA to obtain films and other goodies to make your meetings sparkle. Keep the meetings interesting and informative, and you will have no trouble getting members to attend.

I have formed three model rocket clubs and been instrumental in the formation of several others. In all cases there has always been one serious problem—teaching the newcomers. So in 1965 I embarked upon an experiment that ran for eight years and turned out to be highly successful. I formed the New Canaan (Connecticut) YMCA Space Pioneers Section of the NAR and set up a series of qualifications that every member of the club had to meet. The most important was that every member had to complete a nine-month training course in model rocketry taught by experienced club members. We followed a regular course plan, or syllabus, and used this book in its earlier editions as the text. The training course was intended to "bring everyone up to the same level of ignorance and confusion that other club members enjoyed." It provided each member with the same foundation on which to build his future activities in model rocketry. It eliminated a lot of trial-and-error activity by new members. And it provided a proving ground for many new ideas—the Misfire Alley system, simple beginner's models, simple contests, simplified methods of calculating *CP* and estimated altitudes, and a number of other things that are now commonplace in the hobby of model rocketry.

The training course made use of existing kits available in local hobby stores. Every two weeks there was a forty-five minute lecture about an assigned chapter or subject, often with demonstrations. A flight session was held every two weeks to let the trainees try out the things they had studied. There were regular assignments of outside work involving reading in this book or building models in the individual's home workshop. There were never any workshop sessions where trainees brought their models to the club meeting to work on them together; our meeting time was far too valuable and was used instead to present information to them. The flight sessions were the testing periods in which the trainees proved how well they were getting along. Organized sporting competition was used to keep the work interesting, and often the trainees were pitted against the older members in these contests—and won! Awards, trophies,

Figure 17-4: The New Canaan YMCA Space Pioneers Section of the NAR was a fun group of dads, mothers, sisters, and brothers, all of whom went through the training course that twice took them to national meets to become the top model rocket club in the United States.

and prizes were always given. Some were humorous. Upon completion of the training course, each person was presented with a diploma certifying him to be a "Compleat Model Rocketeere."

How successful was the training program? During every year of the club's existence between 1965 and 1973 when I left it, there was always at least one United States National Champion Model Rocketeer in the club. Twice, the Space Pioneers won the coveted National Championship Section Pennant of the NAR. On the basis of point awards determined by the national rules, the club was never beaten in any local or regional sanctioned contest. One club member became an international FAI medalist in the First World Championships for Space Models in Yugoslavia in 1972.

The training course system of club operation works. This statement is backed up by facts. But if your club tries it, you must carry it through all the way with a great deal of advanced planning. And you can't take shortcuts.

Other than regular flying sessions, your club will be called on to

Figure 17-5: Contest flying with a set of standard rules allows modelers to compare their skills. Scale model competitions where everyone builds the same model from the same set of data have proven extremely popular.

conduct flight demonstrations from time to time. These may be held simply to amaze and impress your friends and school teachers, or they may be more important, such as showing local public safety officials what model rocketry is all about.

When you put on a flight demonstration, do so without an admission charge, even though your club treasury could stand some bolstering. Pass the hat for donations instead, and you will eliminate all sorts of hassles about entertainment taxes and such things.

A demonstration is the time to show off your club shirts, if you have them. It is a good idea to have club shirts because, as your club grows, they will allow you to tell a member from a spectator! This isn't funny; such confusion has happened. Special shirts also serve to identify club members to the RCO when they are in the downrange recovery area. And they help the spectators know who is a member in case a viewer has questions.

It is very important to maintain strict range discipline during a demonstration.

It is also important to fly only models of very high reliability. A demonstration is no place to try out an experimental design. Safety-checking should be extra-strict. A short, successful demonstration is best, starting with two low-altitude models, one with a parachute and the other with a streamer. Follow this with a cloud-buster, a high-performance job with a Type C motor in it. Fly a boost-glider, a two-staged model, and perhaps a large model if the field is big enough. Payload models are always impressive and show that there is more to model rocketry than up-and-down. Fly a camera. Fly an egg, squeezing the flight for all the suspense you can. Save your best stuff until last, and leave the spectators with something good to talk about. *Don't* fly salvos where two or more models are launched simultaneously; that is fireworks-type stuff. Have the PA system going at all times with chatter, telling the spectators what is going on. Use the full countdown ritual and take every opportunity to show your visitors that model rocketry is not a bunch of pyromaniacs playing with fireworks or kids playing with toys, but a serious technical hobby, a technology in miniature.

Remember, model rocketry has brought space technology to Main Street, U.S.A. More people have seen live model rockets launched than have ever watched a live launching in person from the Cape. What they see in person on your model rocket range is, to them, the space program in miniature. Yours may be the only rocket vehicles they have ever seen!

Contests provide lots of the fun in model rocketry. Competition began when one model rocketeer said to another, "My model rocket will go higher than yours!" And the other guy said, "Prove it!"

To keep your club's first contest from becoming *too* hectic and confused, schedule only two very simple events such as spot landing and parachute duration. Little special equipment is required for these events. You should have three judges who are adults and who can render impartial decisions on all of the little protests and complaints that will ensue. Such minor hassles are always part of competition, even with the best aura of sportsmanship. The rules should be understood by every contestant and the operating procedures thoroughly explained before the contest gets under way.

The rules for flight duration events can be quite simple. Ordinarily, the flights are timed by two timers with stopwatches. Timers start their watches when the model takes off and stop the watches when the model goes out of sight, touches the ground, or lands in a tree. Typical flight duration events include: parachute duration, streamer duration, boost-glider duration, rocket glider duration, and even egg-loft duration. There should be a limit on the size of the motor that may be used, but no limit on parachute size, streamer size, or wing area of boost-gliders or rocket gliders. As we have seen in earlier chapters, there are optimum sizes for such items and self-limiting factors that prevent somebody from running away with the event because he used a bigger 'chute, streamer, or wings. Those contestants who have mastered the design factors stand a better chance of winning.

Figure 17-6: Model rocketeers both young and old travel across the United States every year to the National Model Rocket Championships, which is not only the top national meet but also an affair full of camaraderie and sportsmanship where people help each other.

Figure 17-7: The top international model rocket contest is the World Championships, first held in Vrsac, Yugoslavia, in 1972. Here the United States team helps Howard Kuhn load up his scale model of the NASA Javelin sounding rocket.

You can even run simple altitude events using stopwatches and the marker streamer method, as was explained earlier.

If you want to run your meet with standardized national rules, the NAR publishes the United States Model Rocket Sporting Code, a set of rules used on a nationwide basis. All NAR Sections that fly NAR-sanctioned contests must fly by these rules.

As your club grows in contest experience, you will be able to fly more of the advanced and difficult events. For practical operation, however, don't schedule more than five events for an afternoon of flying, depending on the number of contestants.

If your club is an NAR Section, you will get a lot of information on conducting a contest from the NAR. Having all contests run in the same fashion will make it easier for your club when they go to fly in another meet. And it will not take long for your club to be ready to take on another club in an open or regional contest on another field. These can be very fun weekends for everyone.

Then there is the big national meet, the NARAM, held every year by the NAR. Hundreds of model rocketeers from all over North America come to this meet, the oldest continuously held national model rocket competition in the world. The first NARAM was held in Denver, Colorado, in 1959 with twenty-six contestants. Now it has grown to the point where contestants are selected on the basis of their contest record for the year to date, because a NARAM with more than three hundred contestants usually becomes unwieldy. The NARAM has been held at the United States Air Force Academy, the United States Army Aberdeen Proving Ground, NASA Wallops Flight Research Center, and several other locations.

Figure 17-8: America's first international model rocket champions, the United States team at the First World Championships for Space Models: (standing, left to right) Jon Randolph, Shirley Lindgren, Al Lindgren, Jim Worthen; (kneeling, left to right) Bernard Biales, Howard Kuhn, Ellie Stine, and Team Manager Jim Kukowski.

Beyond the NARAM are the international meets and the World Championships for Space Models held every two years under the auspices of the Federation Aeronautique Internationale. The First World Championships were held in Vrsac, Yugoslavia, in 1972 with nine nations competing—Canada, Great Britain, Poland, Czechoslovakia, Bulgaria, Rumania, Egypt, Yugoslavia, and the United States. There is no higher level of competition than the World Championships, the Olympics of model rocketry. To become a World Champion is to be accorded the title of *Weltmeister*, literally "master of the world."

International model rocketry has generated a tremendous amount of international good will and fellowship, and will continue to do so as the years go by.

In addition to competition, there are also national and international model rocket performance records. The national certificates are given by the National Association of Rocketry on behalf of the National Aeronautic Association for model performances that are carefully documented, checked, and certified as being the best performances in their respective categories in the United States. It is even more difficult to set an international model rocket record, but its reward is a huge FAI diploma in French with ribbons and wax seals galore. There are not many of these around. To get one, you have to build a model rocket that is the best in the world.

In club work, competition, and record-setting lie the true fun and advancement in model rocketry. Club work gives you the opportunity to learn how to work with people, something that in the long run may be far more important than knowing about gadgetry. Contests run under accepted rules separate the good modelers from the not-so-good modelers and the best designs from the mediocre ones. Competition sharpens your abilities, capabilities, and mental processes; Charles Darwin pointed out that Mother Nature thinks very highly of competition. It improves things.

Figure 17-9: Ellie Stine, then seventeen, won third place in the parachute duration category at the First World Championships. Her father, the author, was Chief of the FAI Jury.

By now it should be apparent that there is a great deal to this hobby called model rocketry. As many people have learned, it involves nearly all aspects of human endeavor, just like its full-sized counterpart in astronautics. How far do *you* want to go in model rocketry? To the NARAM? To the World Championships? To the moon? To the stars? It's all up to you.

Epilogue

I have been in model rocketry since 1957, and I have been accumulating the technical information and background knowledge for this book ever since. This is the Fourth Edition of *The Handbook of Model Rocketry*. I rewrote it completely from cover to cover because the first three editions became obsolete because of themselves and what they inspired. And I am happy that it happened. If this book does not go out of date, it will be because there has been no progress toward better things in model rocketry.

One key to progress is communication. If you find an error in this book, or if you develop something new that may not be mentioned in these pages, please write to me in care of the publisher.

There are a lot of problems left to be solved in our miniature technology. There are a lot of things left to do. Let's get on with it . . . .

Appendix I

Important Addresses

National Association of Rocketry, P.O. Box 725, New Providence, New Jersey 07974

Centuri Engineering Company, P.O. Box 1988, Phoenix, Arizona 85001

Competition Model Rockets, P.O. Box 7022, Alexandria, Virginia 22307

Estes Industries, Inc., P.O. Box 227, Penrose, Colorado 81240

Flight Systems, Inc., 9300 East Sixty-Eighth Street, Raytown, Missouri 64133

Appendix II

Calculating Model Rocket Flight Performance Using the Acceleration Method

Please refer to the equations on pages 127–137 of Chapter 7.

Assume the following hypothetical model rocket characteristics for this example:

Loaded weight at lift-off: 1.5 ounces = 0.094 pound
Weight at burnout: 1.3 ounces = 0.081 pound
Average thrust of motor: 0.9 pound (4 newtons)
Thrust duration of motor: 1.2 seconds

We compute the lift-off acceleration by using Equation (1) in modified form as follows:

$$A_0 = \left(\frac{F}{W_0} - 1\right) 32.2 = \left(\frac{0.9}{.094} - 1\right) 32.2 = 276.1 \text{ ft/sec/sec}$$

This is the lift-off acceleration. Since the model rocket motor consumes propellant during burning, the model gets lighter as it goes up. Its mass therefore changes, and the acceleration increases. We must therefore compute the burnout acceleration:

$$A_1 = \left(\frac{0.9}{0.081} - 1\right) 32.2 = 325.6 \text{ ft/sec/sec}$$

The model has undergone a change of acceleration, or a surge, during powered flight. To account for the surge in the flight equations requires more than simple algebra, so we must work out a method of simple computing. To do this, we have to compute the average acceleration during powered flight:

$$A_{av} = \frac{A_0 + A_1}{2} = \frac{276.1 + 325.6}{2} = \frac{601.7}{2} = 300.85 \text{ ft/sec/sec}$$

The maximum velocity at burnout can then be computed using Equation (4) where $v_1 = 0$ since the model is starting from rest on the launch pad:

$$V_{max} = A_{av} \, t = 300.85 \times 1.2 = 361 \text{ ft/sec}$$

The average velocity during powered flight is computed using Equation (3):

$$V_{av} = \frac{V_{max}}{2} = \frac{361}{2} = 180.5 \text{ ft/sec}$$

Using Equation (2), we can then compute the burnout altitude:

$$S_{bo} = V_{av} \, t_b = 180.5 \times 1.2 = 216.6 \text{ feet}$$

From this point on, the flight performance calculations follow the identical method shown in the text for computing altitude gained during coasting flight, and total altitude achieved.

Appendix III

Model Rocket *CP* Calculation

From James S. and Judith A. Barrowman

Nose:

L_N = length of nose

For Cone
$(C_N)_N = 2$
$\overline{X}_N = 0.666 L_N$

For Ogive
$(C_N)_N = 2$
$\overline{X}_N = 0.466 L_N$

Conical Transition: (for both increasing and decreasing diameters)

d_F = diameter of front of transition
d_R = diameter of rear of transition
L_T = length of transition piece (distance from d_F to d_R)
X_P = distance from tip of nose to front of transition
d = diameter of base of nose

$$(C_N)_T = 2\left[\left(\frac{d_R}{d}\right)^2 - \left(\frac{d_F}{d}\right)^2\right]$$

Note: $(C_N)_T$ will be negative for conical boattail.

$$\overline{X}_T = X_P + \frac{L_T}{3}\left[1 + \frac{1 - \dfrac{d_F}{d_R}}{1 - \left(\dfrac{d_F}{d_R}\right)^2}\right]$$

Fins: (for multistaged models, calculate each set of fins separately, using a different X_B)

C_R = fin root chord
C_T = fin tip chord
S = fin semispan

L_F = length of fin mid-chord line
R = radius of body rear end
X_R = distance between fin root leading edge and fin tip leading edge parallel to body
X_B = distance from nose tip to fin root chord leading edge

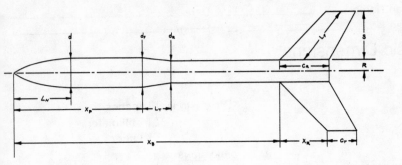

FOR 3 FINS

$$(C_N)_F = \left[1 + \frac{R}{S+R}\right]\left[\frac{12\left(\frac{S}{d}\right)^2}{1 + \sqrt{1 + \left(\frac{2L_F}{C_R+C_T}\right)^2}}\right]$$

FOR 4 FINS

$$(C_N)_F = \left[1 + \frac{R}{S+R}\right]\left[\frac{16\left(\frac{S}{d}\right)^2}{1 + \sqrt{1 + \left(\frac{2L_F}{C_R+C_T}\right)^2}}\right]$$

$$\overline{X}_F = X_B + \frac{X_R}{3}\frac{(C_R+2C_T)}{(C_R+C_T)} + \frac{1}{6}\left[(C_R+C_T) - \frac{(C_RC_T)}{(C_R+C_T)}\right]$$

Total Values:

$$(C_N)_R = (C_N)_N + (C_N)_T + (C_N)_F + \ldots$$

(the sum of the force coefficient C_N of each part calculated)

CP Distance From Nose Tip = $\overline{X}$

$$= \frac{(C_N)_N\,\overline{X}_N + (C_N)_T\overline{X}_T + (C_N)_F\overline{X}_F}{(C_N)_R}$$

(the sum of the products of the force coefficient C_N and the part *CP* of each part divided by the total rocket C_N)

Appendix IV

Rocket Design Charts for Use with Centuri TIR-33

Design Name _____   Date _____

Design No. _____   Designer _____

Basic Dimensions

Dimensions: ☐ Inches
☐ Millimeters
☐ Other _____

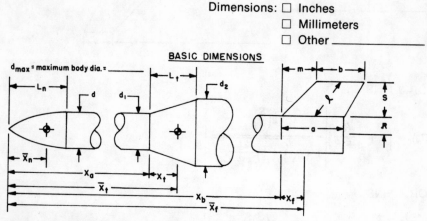

BASIC DIMENSIONS

Note: For conical shoulder, use Chart 1; for conical boattail, use Chart 2.
Dimensional notation same for both.

Sketch of Design

(Basic Dimensions, continued)

GIVEN: Nose L_n = nose length = _____
             d = nose base dia. = _____

		1.	2.
Transition(s)	L_t = transition length =	_____	_____
	d_1 = transition front dia. =	_____	_____
	d_2 = transition rear dia. =	_____	_____
	X_a = nose tip to transition front =	_____	_____

		1.	2.
Fins	n = number of fins =	_____	_____
	a = root chord =	_____	_____
	b = tip chord =	_____	_____
	S = semispan =	_____	_____
	m = leading edge sweep =	_____	_____
	ℓ = mid-chord length =	_____	_____
	R = body rear end radius =	_____	_____
	X_b = nose tip to root leading edge =	_____	_____

TO BE CALCULATED:

$\overline{X}_n$ = CP of nose from nose tip
X_t = transition CP from transition front
$\overline{X}_t$ = transition CP from nose tip
X_f = fin CP from fin root leading edge
$\overline{X}_f$ = fin CP from nose tip
$(C_{n\alpha})_n$ = nose normal force
$(C_{n\alpha})_t$ = transition normal force
$(C_{n\alpha})_{fb}$ = fin-body normal force

d_{max} = maximum body dia. = _____

Design Name _____

Design No. _____

Nose

FORCE: $(C_{n\alpha})_n = 2.00$ (for all shapes)

CP: $L_n =$ _____   ⟵ _____

Shape:

Cone $\overline{X}_n = 0.66\ L_n = 0.66 \times ($ _____ $) =$ _____

Ogive $\overline{X}_n = 0.466\ L_n = 0.466 \times ($ _____ $) =$ _____

Paraboloid $\overline{X}_n = 0.5\ L_n = 0.5 \times ($ _____ $) =$ _____

Design Name _____

Design No. _____

Transition No. _____

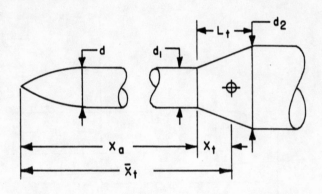

Use for shoulders or boattails

Dimensions:
- d = nose base diameter = _____
- d_1 = transition front dia. = _____
- d_2 = transition rear dia. = _____
- L_t = transition length = _____
- X_a = nose tip to transition front = _____

$$\frac{d_1}{d} = \frac{(\quad\quad\quad\quad)}{(\quad\quad\quad\quad)} = \underline{\quad\quad\quad\quad\quad}$$

$$\frac{d_2}{d_1} = \frac{(\quad\quad\quad\quad)}{(\quad\quad\quad\quad)} = \underline{\quad\quad\quad\quad\quad}$$

FORCE:
For shoulder, use Chart 1.
For boattail, use Chart 2.

$$(C_{n\alpha})_t = \underline{\quad\quad\quad\quad\quad\quad\quad\quad} \longleftarrow \underline{\quad\quad}$$

CP: From Chart 3:

$$\frac{X_t}{L_t} = \frac{\Delta X_c}{L} = \underline{\quad\quad\quad\quad\quad}$$

$$X_t = \left(\frac{X_t}{L_t}\right) \times L_t$$
$$= (\quad\quad\quad) \times (\quad\quad\quad) = \underline{\quad\quad\quad\quad\quad}$$

$$\bar{X}_t = X_a + X_t$$
$$= (\quad\quad\quad) + (\quad\quad\quad) = \underline{\quad\quad\quad\quad\quad} \longleftarrow$$

Design Name _____

Design No. _____

Fins No. _____

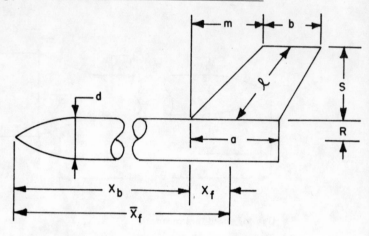

Dimensions: n = number of fins = _____
               a = root chord = _____
               b = tip chord = _____
               S = semispan = _____
               m = leading edge sweep = _____
               ℓ = mid-chord line length = _____
               R = radius of body rear end = _____
               X_b = nose tip to root leading edge = _____

FORCE: $\dfrac{S}{d} = \dfrac{(\qquad\qquad)}{(\qquad\qquad)} =$ _____

          $\dfrac{\ell}{a+b} = \dfrac{(\qquad\qquad)}{(\quad)+(\quad)} = \dfrac{(\qquad\qquad)}{(\qquad\qquad)} =$ _____

FOR ELLIPTICAL FINS:

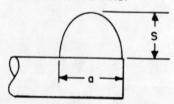

$\dfrac{\ell}{a+b} = \dfrac{S}{a} \times .638 = \dfrac{(\qquad\qquad)}{(\qquad\qquad)} \times .638 = (\qquad\qquad) \times .638 =$ _____

(Fins, continued)

From Chart 4: $(C_{n\alpha})_f =$ _____ (for $n = 4$)

$(\quad\quad\quad) \times .75 =$ _____ (for $n = 3$)

$(\quad\quad\quad) \times 1.5 =$ _____ (for $n = 6$)

$$\frac{R}{S} = \frac{(\quad\quad\quad)}{(\quad\quad\quad)} = \underline{\hspace{3cm}}$$

From Chart 5: $K_{fb} =$ _____

$(C_{n\alpha})_{fb} = (C_{n\alpha})_f \times K_{fb} = (\quad\quad) \times (\quad\quad) =$ _____ $\longleftarrow$

CP: $$\frac{m}{a} = \frac{(\quad\quad\quad)}{(\quad\quad\quad)} = \underline{\hspace{3cm}}$$

$$\frac{b}{a} = \frac{(\quad\quad\quad)}{(\quad\quad\quad)} = \underline{\hspace{3cm}}$$

From Chart 6: $$\frac{X_f}{a} = \frac{\Delta X_f}{a} = \underline{\hspace{3cm}}$$

$$X_f = \frac{X_f}{a} \times a = \frac{(\quad\quad\quad)}{(\quad\quad\quad)} \times (\quad\quad\quad) = \underline{\hspace{3cm}}$$

$$\overline{X}_f = X_b + X_f = (\quad\quad) + (\quad\quad) = \underline{\hspace{3cm}} \longleftarrow$$

FOR ELLIPTICAL FINS

$X_f = a \times .288 = (\quad\quad\quad) \times .288 =$ _____

CP-CG Tables

Design Name _____

Center of Pressure

Center of Gravity

Part	Force ($C_{n\alpha}$)	Part CP X	Moment ($C_{n\alpha}$) × $\overline{X}$	Weight (W)	Part CG from nose tip $\overline{Y}$	Moment (W) × $\overline{Y}$
Nose	2.00					
Payload						
Body						
Recovery						
Transition 1						
Transition 2						
Fins 1						
Fins 2						
Motor						
	Total Force $(C_{n\alpha})_R =$		Total Moment $M_R =$	Total Weight $W_0 =$		Total Moment $M_W =$

(CP-CG Tables, continued)

Design CP $= \bar{X} = \dfrac{\text{Total Moment}}{\text{Total Force}} = \dfrac{M_R}{(C_{n\alpha})_R} = \dfrac{(\quad)}{(\quad)} = \underline{\hspace{3cm}}$ **Aft of nose tip**

Design CG (calculated) $= \bar{G} = \dfrac{\text{Total Moment}}{\text{Total Weight}} = \dfrac{M_w}{W_0} = \dfrac{(\quad)}{(\quad)} = \underline{\hspace{3cm}}$ **Aft of nose tip**

or

Design CG (measured) $= \bar{G} = \underline{\hspace{3cm}}$ **Aft of nose tip**

Stability: CP $-$ **CG** $= \bar{X} - \bar{G} = (\quad) - (\quad) = \underline{\hspace{3cm}}$ **(+ = OK, − = No Go)**

Stability Margin: $\dfrac{\text{Stability}}{\text{max. body dia.}} = \dfrac{\bar{X} - \bar{G}}{d_{max}} = \dfrac{(\quad)}{(\quad)} = \underline{\hspace{3cm}}$ **(more than I = OK, less than I = No Go)**

Appendix V

Three-station Elevation-only Tracking System

Please refer to Figure 15-9 on page 296. These tables have been computed for three stations located in a straight line with the two end trackers located 100 meters (328.1 feet) from the middle tracker, giving a total base line length of 200 meters (656.2 feet).

STEP 1
From Table 1:

End Tracker No. 1 elevation angle: _____   End Tracker Number, Table 1: _____
End Tracker No. 2 elevation angle: _____   End Tracker Number, Table 1: _____
                                            Add Together: _____
Middle Tracker elevation angle: _____   Middle Tracker Number: _____
                                            Subtract: _____

TABLE 1

Angle	End Tracker	Middle Tracker	Angle	End Tracker	Middle Tracker
1	3282.139	6564.279	26	4.203	8.407
2	820.034	1640.069	27	3.851	7.703
3	364.089	728.179	28	3.537	7.074
4	204.509	409.018	29	3.254	6.509
5	130.646	261.292	30	2.999	5.999
6	90.523	181.046	31	2.769	5.539
7	66.330	132.660	32	2.561	5.122
8	50.628	101.256	33	2.371	4.742
9	39.863	79.726	34	2.197	4.395
10	32.163	64.326	35	2.039	4.079
11	26.466	52.932	36	1.894	3.788
12	22.133	44.267	37	1.761	3.522
13	18.761	37.523	38	1.638	3.276
14	16.086	32.172	39	1.524	3.049
15	13.928	27.856	40	1.420	2.840
16	12.162	24.324	41	1.323	2.646
17	10.698	21.396	42	1.233	2.466
18	9.472	18.944	43	1.149	2.299
19	8.434	16.868	44	1.072	2.144
20	7.548	15.097	45	0.999	1.999
21	6.786	13.572	46	0.932	1.865
22	6.126	12.252	47	0.869	1.739
23	5.550	11.100	48	0.810	1.621
24	5.044	10.089	49	0.755	1.511
25	4.598	9.197	50	0.704	1.408

Angle	End Tracker	Middle Tracker	Angle	End Tracker	Middle Tracker
51	0.655	1.311	71	0.118	0.237
52	0.610	1.220	72	0.105	0.211
53	0.567	1.135	73	0.093	0.186
54	0.527	1.055	74	0.082	0.164
55	0.490	0.980	75	0.071	0.143
56	0.454	0.909	76	0.062	0.124
57	0.421	0.843	77	0.053	0.106
58	0.390	0.780	78	0.045	0.090
59	0.361	0.722	79	0.037	0.075
60	0.333	0.666	80	0.031	0.062
61	0.307	0.614	81	0.025	0.050
62	0.282	0.565	82	0.019	0.039
63	0.259	0.519	83	0.015	0.030
64	0.237	0.475	84	0.011	0.022
65	0.217	0.434	85	0.007	0.015
66	0.198	0.396	86	0.004	0.009
67	0.180	0.360	87	0.002	0.005
68	0.163	0.326	88	0.001	0.002
69	0.147	0.294	89	0.000	0.000
70	0.132	0.264	90	0.000	0.000

STEP 2

From Table 2:

Look up the "Subtract" number obtained above in the "Sum of Values" column of Table 2. "Height" number opposite "Sum of Values" number is the achieved altitude in meters.

Sum of Values: _____ Height: _____ meters.

TABLE 2

Sum of Values	Height	Sum of Values	Height	Sum of Values	Height	Sum of Values	Height
20000.000	1	165.289	11	45.351	21	20.811	31
5000.000	2	138.888	12	41.322	22	19.531	32
2222.222	3	118.343	13	37.807	23	18.365	33
1250.000	4	102.040	14	34.722	24	17.301	34
800.000	5	88.888	15	32.000	25	16.326	35
555.555	6	78.125	16	29.585	26	15.432	36
408.163	7	69.204	17	27.434	27	14.609	37
312.500	8	61.728	18	25.510	28	13.850	38
246.913	9	55.401	19	23.781	29	13.149	39
200.000	10	50.000	20	22.222	30	12.500	40

(Table 2, continued)

Sum of Values	Height	Sum of Values	Height	Sum of Values	Height	Sum of Values	Height	Sum of Values	Height
11.897	41	2.415	91	1.005	141	0.548	191	0.344	241
11.337	42	2.362	92	0.991	142	0.542	192	0.341	242
10.816	43	2.312	93	0.978	143	0.536	193	0.338	243
10.330	44	2.263	94	0.964	144	0.531	194	0.335	244
9.876	45	2.216	95	0.951	145	0.525	195	0.333	245
9.451	46	2.170	96	0.938	146	0.520	196	0.330	246
9.053	47	2.125	97	0.925	147	0.515	197	0.327	247
8.680	48	2.082	98	0.913	148	0.510	198	0.325	248
8.329	49	2.040	99	0.900	149	0.505	199	0.322	249
8.000	50	2.000	100	0.888	150	0.500	200	0.320	250
7.689	51	1.960	101	0.877	151	0.495	201	0.317	251
7.396	52	1.922	102	0.865	152	0.490	202	0.314	252
7.119	53	1.885	103	0.854	153	0.485	203	0.312	253
6.858	54	1.849	104	0.843	154	0.480	204	0.310	254
6.611	55	1.814	105	0.832	155	0.475	205	0.307	255
6.377	56	1.779	106	0.821	156	0.471	206	0.305	256
6.155	57	1.746	107	0.811	157	0.466	207	0.302	257
5.945	58	1.714	108	0.801	158	0.462	208	0.300	258
5.745	59	1.683	109	0.791	159	0.457	209	0.298	259
5.555	60	1.652	110	0.781	160	0.453	210	0.295	260
5.374	61	1.623	111	0.771	161	0.449	211	0.293	261
5.202	62	1.594	112	0.762	162	0.444	212	0.291	262
5.039	63	1.566	113	0.752	163	0.440	213	0.289	263
4.882	64	1.538	114	0.743	164	0.436	214	0.286	264
4.733	65	1.512	115	0.734	165	0.432	215	0.284	265
4.591	66	1.486	116	0.725	166	0.428	216	0.282	266
4.455	67	1.461	117	0.717	167	0.424	217	0.280	267
4.325	68	1.436	118	0.708	168	0.420	218	0.278	268
4.200	69	1.412	119	0.700	169	0.417	219	0.276	269
4.081	70	1.388	120	0.692	170	0.413	220	0.274	270
3.967	71	1.366	121	0.683	171	0.409	221	0.272	271
3.858	72	1.343	122	0.676	172	0.405	222	0.270	272
3.753	73	1.321	123	0.668	173	0.402	223	0.268	273
3.652	74	1.300	124	0.660	174	0.398	224	0.266	274
3.555	75	1.280	125	0.653	175	0.395	225	0.264	275
3.462	76	1.259	126	0.645	176	0.391	226	0.262	276
3.373	77	1.240	127	0.638	177	0.388	227	0.260	277
3.287	78	1.220	128	0.631	178	0.384	228	0.258	278
3.204	79	1.201	129	0.624	179	0.381	229	0.256	279
3.125	80	1.183	130	0.617	180	0.378	230	0.255	280
3.048	81	1.165	131	0.610	181	0.374	231	0.253	281
2.974	82	1.147	132	0.603	182	0.371	232	0.251	282
2.903	83	1.130	133	0.597	183	0.368	233	0.249	283
2.834	84	1.113	134	0.590	184	0.365	234	0.247	284
2.768	85	1.097	135	0.584	185	0.362	235	0.246	285
2.704	86	1.081	136	0.578	186	0.359	236	0.244	286
2.642	87	1.065	137	0.571	187	0.356	237	0.242	287
2.582	88	1.050	138	0.565	188	0.353	238	0.241	288
2.524	89	1.035	139	0.559	189	0.350	239	0.239	289
2.469	90	1.020	140	0.554	190	0.347	240	0.237	290

Sum of Values	Height	Sum of Values	Height	Sum of Values	Height	Sum of Values	Height	Sum of Values	Height
0.236	291	0.171	341	0.130	391	0.102	441	0.082	491
0.234	292	0.170	342	0.130	392	0.102	442	0.082	492
0.232	293	0.169	343	0.129	393	0.101	443	0.082	493
0.231	294	0.169	344	0.128	394	0.101	444	0.081	494
0.229	295	0.168	345	0.128	395	0.100	445	0.081	495
0.228	296	0.167	346	0.127	396	0.100	446	0.081	496
0.226	297	0.166	347	0.126	397	0.100	447	0.080	497
0.225	298	0.165	348	0.126	398	0.099	448	0.080	498
0.223	299	0.164	349	0.125	399	0.099	449	0.080	499
0.222	300	0.163	350	0.125	400	0.098	450	0.080	500
0.220	301	0.162	351	0.124	401	0.098	451	0.079	501
0.219	302	0.161	352	0.123	402	0.097	452	0.079	502
0.217	303	0.160	353	0.123	403	0.097	453	0.079	503
0.216	304	0.159	354	0.122	404	0.097	454	0.078	504
0.214	305	0.158	355	0.121	405	0.096	455	0.078	505
0.213	306	0.157	356	0.121	406	0.096	456	0.078	506
0.212	307	0.156	357	0.120	407	0.095	457	0.077	507
0.210	308	0.156	358	0.120	408	0.095	458	0.077	508
0.209	209	0.155	359	0.119	409	0.094	459	0.077	509
0.208	310	0.154	360	0.118	410	0.094	460	0.076	510
0.206	311	0.153	361	0.118	411	0.094	461	0.076	511
0.205	312	0.152	362	0.117	412	0.093	462	0.076	512
0.204	313	0.151	363	0.117	413	0.093	463	0.075	513
0.202	314	0.150	364	0.116	414	0.092	464	0.075	514
0.201	315	0.150	365	0.116	415	0.092	465	0.075	515
0.200	316	0.149	366	0.115	416	0.092	466	0.075	516
0.199	317	0.148	367	0.115	417	0.091	467	0.074	517
0.197	318	0.147	368	0.114	418	0.091	468	0.074	518
0.196	319	0.146	369	0.113	419	0.090	469	0.074	519
0.195	320	0.146	370	0.113	420	0.090	470	0.073	520
0.194	321	0.145	371	0.112	421	0.090	471	0.073	521
0.192	322	0.144	372	0.112	422	0.089	472	0.073	522
0.191	323	0.143	373	0.111	423	0.089	473	0.073	523
0.190	324	0.142	374	0.111	424	0.089	474	0.072	524
0.189	325	0.142	375	0.110	425	0.088	475	0.072	525
0.188	326	0.141	376	0.110	426	0.088	476		
0.187	327	0.140	377	0.109	427	0.087	477		
0.185	328	0.139	378	0.109	428	0.087	478		
0.184	329	0.139	379	0.108	429	0.087	479		
0.183	330	0.138	380	0.108	430	0.086	480		
0.182	331	0.137	381	0.107	431	0.086	481		
0.181	332	0.137	382	0.107	432	0.086	482		
0.180	333	0.136	383	0.106	433	0.085	483		
0.179	334	0.135	384	0.106	434	0.085	484		
0.178	335	0.134	385	0.105	435	0.085	485		
0.177	336	0.134	386	0.105	436	0.084	486		
0.176	337	0.133	387	0.104	437	0.084	487		
0.175	338	0.132	388	0.104	438	0.083	488		
1.174	339	0.132	389	0.103	439	0.083	489		
0.173	340	0.131	390	0.103	440	0.083	490		

Appendix VI

Sample NAR Section Bylaws

These are sample bylaws. Individual Sections may wish to alter them or add to them, due to local circumstances. Please *do not* merely fill in the blanks of this sheet and forward it to NAR Headquarters for approval; make your own copies, and be sure you have enough for your members and a copy for NAR Headquarters. The purpose of these sample bylaws is to provide a guide for each group in drawing up its own bylaws. All bylaws and amendments thereto must be approved in writing by NAR Headquarters.

Article 1, Name: The name of this organization shall be the _____ Section of the National Association of Rocketry.

Article 2, Purpose: It shall be the purpose of this Section to (a) aid and abet the aims and purposes of the NAR in _____ (locale) _____, (b) to operate and maintain a model rocket range in accordance with the NAR Standards and Regulations, (c) to hold meetings for the purpose of aiding and encouraging all those interested in rocketry, and (d) to engage in other scientific, educational, or related activities as the NAR, the Section, or the Section Board of Directors may from time to time deem necessary or desirable in connection with the foregoing.

Article 3, Membership: All members of this Section shall be NAR members in good standing who reside in _____ (locale) _____.

Article 4, Dues: Dues shall be $_____ per year, payable in advance. These Section dues are separate and distinct from national dues paid to the NAR. All dues monies shall be kept in a General Fund by the Secretary-Treasurer and shall be paid out by him only on order of the Section Board of Directors. Special assessments may be levied by a majority vote of the members present and voting at any meeting of the Section, provided notice of such intent is given in writing to each member at least five days preceding such a meeting.

Article 5, Meetings: Meetings of the Section shall be held at least _____ _____ times per year at times and places designated by the Section Board of Directors. Operation of the rocket range shall not be considered a meeting. A quorum shall consist of 50% of the membership of the Section. Meetings shall be conducted and governed by *Roberts' Rules of Order, Revised.*

Article 6, Board of Directors: The Board of Directors of this Section shall consist of the three officers, one member at large, and a Senior Member of the NAR, who shall be designated by the NAR as Section Advisor.

Article 7, Officers: The officers of this Section shall consist of a President, a Vice-President, and a Secretary-Treasurer, all of whom shall be members of the Section and of the NAR.

Article 8, Elections: Elections of officers and members of the Board of Directors shall take place at the first meeting of the calendar year. All officers and members of the Board shall serve a term of one year. Vacancies in offices and on the Board shall be filled by nomination and election of a Section member to fill the unexpired term of office and shall take place at the Section meeting at which the vacancy is announced. Nominations for all elections shall be made from the floor, and the candidate having the largest number of votes shall be elected.

Article 9, Committees: There shall be three Standing Committees of the Section, plus such additional committees as the Board of Directors may from time to time deem necessary or desirable. The Standing Committees are as follows:

(a) Operations Committee shall be in charge of the Section's model rocket range, shall monitor the experimental technical activities of the Section members, and shall act as safety inspectors. The Chairman of this Committee shall be a Senior Member of the NAR in good standing and shall act as Range Safety and Control Officer under the NAR Official Standards and Regulations.

(b) The Contests and Records Committee shall be in charge of all arrangements for contests and shall monitor all national-record attempts by Section members. The Committee shall contain at least one Leader Member of the NAR.

(c) The Activities Committee shall be in charge of making all arrangements for all Section meetings, for conducting membership campaigns, and for carrying on public relations.

(d) The Section President shall be an ex-officio member of all committees.

Article 10, Amendments: These bylaws may be amended by a two-thirds vote of those Section members present and voting at any meeting of the Section, providing written notice of the pending amendment has been sent to the membership of the Section at least five days in advance of such meeting. No amendment of these bylaws shall be in force until approved by NAR Headquarters.

Adopted: _____

Approved by NAR Headquarters: _____

_____

Bibliography

Abbott, Ira H., and Von Doenhoff, Albert E., *Theory of Wing Sections.* New York, Dover Publications, Inc., 1959.

Aerodynamics for Naval Aviators, NavWeps 00-80T-80. Washington, D.C. 20025, Superintendent of Documents, Government Printing Office, 1960.

American Radio Relay League, Inc., *The Radio Amateur's Handbook.* West Hartford, Connecticut, 1974.

Barrowman, James S., *Calculating the Center of Pressure of a Model Rocket,* Technical Information Report TIR-33. Phoenix, Arizona, Centuri Engineering Company, 1968.

————, *Stability of a Model Rocket in Flight,* Technical Information Report TIR-30. Phoenix, Arizona, Centuri Engineering Company, 1968.

Davis, L. et al., *Exterior Ballistics of Rockets.* New York and Princeton, New Jersey, D. Van Nostrand Company, Inc., 1958.

Exploring in Aerospace Rocketry, NASA TM X-52387. Cleveland, Ohio, NASA Lewis Research Center, 1968.

Gregorek, Dr. Gerald M., *Aerodynamic Drag of Model Rockets,* Estes Model Rocket Technical Report TR-11. Penrose, Colorado, Estes Industries, Inc., 1970.

Hertz, Louis H., *The Complete Book of Model Aircraft, Spacecraft, and Rockets.* New York, Crown Publishers, Inc., 1967.

Hobbs, Marvin, *Basics of Missile Guidance and Space Techniques.* New York, John F. Rider Publisher, Inc., 1960.

————, *Fundamentals of Rockets, Missiles, and Spacecraft.* New York, John F. Rider Publisher, Inc., 1962.

Hoerner, Dr. Sighard F., *Fluid Dynamic Drag.* 148 Busteeds Drive, Midland Park, New Jersey, Dr. Sighard F. Hoerner, 1958.

(Bibliography, continued)

Hoffman, R. J., *Model Aeronautics Made Painless*. Halesite, New York, Timely Publications, 1964.

Hunter, Maxwell W., II, *Thrust Into Space*. New York, Holt, Rinehart, and Winston, Inc., 1966.

Ley, Willy, *Rockets, Missiles and Men in Space*. New York, The Viking Press, 1968.

Lowry, Peter, and Griffith, Field, *Model Rocketry, Hobby of Tomorrow*. Garden City, New York, Doubleday and Company, Inc., 1971.

Malewicki, Douglas J., *Model Rocket Altitude Performance*, Technical Information Report TIR-100. Phoenix, Arizona, Centuri Engineering Company, 1968.

Mandell, Gordon K. et al., *Topics in Advanced Model Rocketry*. Cambridge, Massachusetts, and London, England, The MIT Press, 1973.

McEntee, Howard G., *The Model Aircraft Handbook*. New York, Thomas Y. Crowell Company, 1968.

Millikan, Dr. Clark B., *Aerodynamics of the Airplane*. New York, John Wiley and Sons, Inc., 1941.

Model Rocket Altitude Prediction Charts, Estes Model Rocket Technical Report TR-10. Penrose, Colorado, Estes Industries, Inc., 1970.

Moulton, R. G., *Aeromodeller Pocket Data Book*. 13-35 Bridge Street, Hemel Hempstead, Hertford, England, The Model Aeronautical Press, Ltd., 1961.

————, *Design for Aeromodellers*. 13-35 Bridge Street, Hemel Hempstead, Hertford, England, The Model Aeronautical Press, Ltd., 1963.

Shapiro, Ascher H., *Shape and Flow*. Garden City, New York, Doubleday and Company, Inc., 1961.

Stine, G. Harry, *The Model Rocketry Manual*. New York, Sentinel Books Publishers, Inc., 1970.

Wiech, Raymond E., and Strauss, Robert F., *Fundamentals of Rocket Propulsion*. New York, Reinhold Publishing Corporation, 1960.

Zaic, Frank, *Circular Airflow and Model Aircraft*. Northridge, California, Model Aeronautical Publications, 1964.

Index of Tables

Index